WATER QUALITY MODELING

Volume III
Application to Estuaries

Wu-Seng Lung, Ph.D., P.E.

Associate Professor
School of Engineering and
Applied Science
Department of Civil Engineering
University of Virginia
Charlottesville, Virginia

CRC

CRC Press
Boca Raton Ann Arbor London Tokyo

Library of Congress Cataloging-in-Publication Data

(Revised for vol. 3)

Water quality modeling.

Includes bibliographical references and index.
Contents: v. 1. Transport and surface exchange in rivers / author, Steve C. McCutcheon— — 3. Application to estuaries / by Wu-Seng Lung— v. 4. Decision support techniques for lakes and reservoirs / editor, Brian Henderson-Sellers.
1. Water quality—Mathematical models. 2. Water chemistry—mathematical models. I. French, Richard H.

TD370.W3955 1989 628.1′61
ISBN 0-8493-6971-1 (v. 1)
ISBN 0-8493-6973-8 (v. 3)
for Library of Congress 89-25333
CIP

Direct all inquiries to CRC Press, Inc., 2000 Corporate Blvd., N. W., Boca Raton, Florida, 33431.

International Standard Book Number 0-8493-6973-8

Library of Congress Card Number 89-25333
Printed in the United States 1 2 3 4 5 6 7 8 9 0

Printed on acid-free paper

DEDICATION

To my wife, Kathy

AUTHOR

Wu-Seng Lung, Ph.D., P.E., received his M.S. degree in Hydrology/Hydraulics from the University of Minnesota and Ph.D. degree in Environmental Engineering from the University of Michigan. Between 1975 and 1983, he worked at three environmental consulting firms applying water quality modeling to various projects for regulatory agencies, industrial clients, trade associations, and law firms on numerous environmental studies. Since 1983, Dr. Lung has been on the faculty of the Civil Engineering Department at the University of Virginia and is now an Associate Professor.

He has over 18 years of experience in modeling of natural waters, including streams, rivers, lakes, reservoirs, and estuaries for a variety of water quality problems. In 1991, he was named by the U.S. EPA to a review panel for the EPA Chesapeake Bay Program, providing guidance to future water quality modeling work on the Bay watershed. He also contributed to the recent revision of the EPA Simplified Method for conducting WLA for small POTWs discharging into low-flow streams. Dr. Lung was the major contributor to EPA's 1992 revision of the WLA manual for stream and river BOD/DO and nutrients. He was selected to the EPA Surface Water Assessment Technical Team (SWAT) to provide technical assistance to the development of TMDL (total maximum daily load) for point and nonpoint source control.

Dr. Lung has served as consultant on water quality modeling to a number of organizations including U.S. EPA, Army Engineers Waterways Experiment Station, state regulatory agencies, consulting firms, and industries.

PREFACE

Water quality modeling, particularly for estuarine systems, has advanced rapidly during the past two decades in response to the need for managing important water resources. The planning and preparation of a series of volumes on water quality modeling was a timely task. When the writing of this volume on estuarine water quality modeling was proposed, I had mixed reactions. Initially, I felt there were quite a few documents, although fragmented, available to the profession. However, following a more careful review of these documents during the preparation of my proposal for the volume, I realized that a need existed for a well synthesized volume on this topic – particularly from a practitioner's point of view. The contents of this volume reflect this theme. Thus, many case studies of prototype estuaries are presented to provide a perspective of estuarine water quality modeling for management, and not just for the sake of modeling.

This volume is based on lecture material given to graduate students at the University of Virginia. Other material is taken from the author's research work and consulting practice ranging from developing a modeling framework to applications of existing water quality models on site specific studies. This volume provides a practical approach to the various facets of the subject and emphasized the application of estuarine models in formulating water quality management strategies. The style of this volume is consistent with that of other volumes in the series, i.e., serving as a useful reference to the profession. It is not intended to be *the* book for estuarine water quality modeling. Yet, it offers much information, from the case studies, on a variety of subtle technical issues which would not be found in other sources.

One message of this volume is that water quality modeling does not generate data. Rater, models are used to interpret data from the prototype. Because of this nature, water quality data are vital to the success of any modeling effort. How to interpret water quality data through correct use of models, is another objective of this volume.

This preface would not be complete without special thanks to my teachers. First, Raymond P. Canale of the University of Michigan introduced me to this profession. My subsequent years at Hydroscience, Inc. provided me with much practical experience on water quality modeling under the guidance of Robert V. Thomann and Donald J. O'Connor at Manhattan College. I still enjoy my working experience with Robert V. Thomann who taught me much on the Lake Ontario modeling study for the International Joint Commission. Perhaps the most profound impact on my professional career comes from working with Donald J. O'Connor with whom I spent my last two years at Hydroscience. Needless to say, I still value the interesting work we did together, and most importantly, the wonderful world of estuaries to which he introduced me. These individuals are my role models, leading me to a relentless pursuit of excellence.

Finally, I would like to thank Steve C. McCutcheon with the U.S. EPA, my colleague and friend, who invited me to participate in this series. He read the preliminary manuscript and offered many suggestions that have made this volume better. Without his continuing encouragement and counsel, I would not have completed this volume – albeit a few years behind schedule.

TABLE OF CONTENTS

Chapter 1

INTRODUCTION TO ESTUARINE WATER QUALITY MODELING

I. INTRODUCTION

In Volume I of this series: *Transport and Surface Exchange in Rivers*,[1] McCutcheon stated that models are necessary to both describe and predict water quality conditions. The same statement applies to estuaries. To put it into a perspective and to appreciate the need for estuarine water quality models, it is necessary to examine the water quality pollution control approaches that have been tried in the U.S. for the past two decades.

Before 1972, water pollution control efforts were based on achievement of ambient water quality standards.[2] Although this approach was economically efficient, it proved virtually impossible to administer, because of the difficulties in translating ambient standards into end-of-pipe effluent limits for individual dischargers, wasteload allocations (WLA). The result was regulatory frustration and very little cleanup. One of the factors that contributed to the frustration was that the "translating" technology, water quality modeling, was not ready at that time.

In 1972, the Federal Water Pollution Control Act was amended to require, at a minimum, a level of control based on available treatment technology. Under this approach, the need to determine impact on the ambient environment was largely eliminated. Instead, regulators had to determine appropriate control technology which is a much simpler technical task. The result was substantial reduction in pollution, even if this was achieved in areas in which cleanup was not needed to meet water quality goals. This was technology-based approach to water pollution control. Over the next half dozen years, secondary treatment was promulgated as the minimum level for all POTWs (publicly owned treatment works) on the assumption that the expected water quality responses were worth the expenditure. Wasteload allocation was used specifically for those instances where there was some doubt that the water quality standards could be achieved by secondary treatment alone.

The pendulum continued to swing between wasteload allocation and effluent requirements.[3] In 1977, the Clean Water Act was passed. Water quality criteria for toxic substances were prepared and, in 1979, the EPA indicated that any requests for construction grants for advanced waste treatment must be rigorously justified through a cost-benefit analysis. It was clear by 1982 that EPA strongly favored water quality-based effluent limitations, rather than technology-based limits, as a basic water quality pollution control strategy, thereby reversing the trend. By that time, water quality modeling technology had advanced significantly and was ready to address a variety of water quality problems. This history indicates a clearer recognition of the importance of a rational approach to water quality management has finally emerged. Thus, contemporary water quality management programs are an integrated system of basic effluent requirements, supplemented by specific analyses of individual situations to arrive at a meaningful allowable discharge.

In the 1990s, wasteload allocation continues to be an important part of this overall process.[3] In fact, the Water Quality Planning and Management Regulation (40 CFR 130) links a number of Clean Water Act sections, including sections 303(d), to form the water quality-based approach to protecting and cleaning up the nation's waters. One of permit writers' goals is to determine what effluent composition will protect aquatic organisms and human health. Exposure assessment includes defining how much of the waterbody is subject to the exceedance of criteria, for how long, and how frequently. The first step is to evaluate the effluent plume dispersion. If mixing is not rapid and complete, and if state water quality standards allow a mixing zone, the wasteload allocation should include a mixing zone analysis.

TABLE 1
History of Water Pollution Control and Water Quality Modeling Applications

Year	Water pollution control	Water quality modeling applications
Pre-1970	To achieve ambient water quality standards	
	Impossible to administer (translating ambient standards into end-of-pipe effluent limits)	BOD/DO modeling
1970	EPA established	Aquatic plants/nutrient Eutrophication modeling
1972	Federal Water Pollution Control Act amended	Hydrothermal modeling
	NPDES permit program for municipal and industrial discharges (technology-based control)	
	Substantial reduction in pollution	
	Allowing variances	
1977	Clean Water Act passed	Toxics modeling
	Water quality criteria for toxic substances	
1982	Water quality-based control	Wasteload allocation modeling
	Water quality-based toxics control	Fate and transport modeling
	Superfund	Sediment modeling
1990s	Clean Water Act reauthorization	Total maximum daily load (TMDL) modeling

Finally, Section 130.7 describes the TMDL (total maximum daily load) process. TMDL is the sum of the individual WLAs for point sources and LAs for nonpoint sources and natural background.[4] Thus, the need for WLAs and water quality modeling has never been greater. A historical summary of water pollution control approaches and the development and applications of water quality models is presented in Table 1.

II. ESTUARIES

Estuaries are coastal water bodies where freshwater meets the sea. They are traditionally defined as semi-enclosed bodies of water having a free connection with the open sea and within which sea water is measurably diluted with freshwater entering from land drainage.[5] These classical estuaries are the lower reaches of rivers where saline and freshwater mix due to tidal action. This definition has been extended to include inland bays and lakes that receive riverine discharge. For example, the backwater river reaches draining into the Great Lakes are considered estuaries.

The seaward end of an estuary is easily defined because it is connected to the sea. The landward end, however, is not that well defined. Generally, tidal influence in a river system extends further inland than salt intrusion. That is, the water close to the fall line of the estuary may not be saline, but it may still be tidal. Thus, the estuary is limited by the requirement that both salt and freshwater be measurably present. The exact location of the salt intrusion depends on the freshwater flow rate which can vary substantially from one season to another. This definition also separates estuaries from coastal bays (embayments) by the requirement for a freshwater inflow and measurable salt water dilution.[6]

Under a topographical system, estuaries are divided into four subclasses:

- Drowned river valley (coastal plain estuary)
- Fjord-like estuaries
- Bar-built estuaries
- Tectonic process estuaries

TABLE 2
Estuarine Water Quality Problems

Water quality constituent	Water quality impact
Salinity (dissolved solids)	Alteration of local salinity regime through dilution
Suspended solids	Altering the habitat of benthic organisms
	Serving as a carrier of contaminants
Bacteria and viruses	In runoff from farms and feedlots
	Effluents from municipal and industrial wastewater discharges
	Pathogens may be transported to shellfish habitat
Dissolved oxygen	Requirement for most aquatic organisms
	Seasonal or diurnal depletion of DO disrupts or displaces estuarine communities
	Best conventional indicator of water quality problems
Nutrients	Excessive nutrient loading can stimulate overproduction of some species of algae
	Periodic phytoplankton blooms can cause widely fluctuating DO levels and DO depletion in benthic and downstream areas
Toxic substances	High concentrations of ammonia, metals, and many organic chemicals can disable or kill aquatic organisms
	Acute toxicity is caused by high exposure to pollutants for short periods of time

Most of the estuaries in the U.S. fall into the drowned river class. Another classification system is based on physical processes and the parameters used in physical classification are directly applicable to estuarine water quality analysis. That is, based on the velocity and salinity patterns in the water column, estuaries can be divided into three classes:

- Stratified (salt wedge) estuary (e.g., the Mississippi River Estuary)
- Well-mixed (e.g., the Delaware and Raritan River estuaries)
- Partially mixed (e.g., the James River Estuary in Virginia)

There are a number of reasons that make estuaries an important natural water resource. First, estuaries are biologically productive. They are the spawning and nursery grounds for many important aquatic biota, including coastal fish and invertebrates. Second, they serve as receiving waters for wastewater discharges. Many major cities and ports are located on estuaries, affecting their quality through domestic and industrial wastewater and dredging. For example, about 25 to 30% of all U.S. publicly owned treatment works (POTW) discharges enter estuaries and coastal waters. Third, many estuaries are important waterways for navigation use.

III. ESTUARINE WATER QUALITY PROBLEMS

A number of water quality problems have been observed in estuaries. Table 2 lists some well-documented water quality problems. A well-known example of estuarine water quality management using water quality models is the Chesapeake Bay, the largest estuary in North America. A time history of eutrophication control of the Chesapeake Bay is summarized in Table 3.

As explained in Chapter 2, different water quality problems are characterized by different spatial and temporal scales. As such, selecting or developing appropriate models is an important issue in practice. The evolution of a three-dimensional (3-D), time-variable water quality model followed a successful modeling study in which a steady-state water quality model was developed and served as the basis for recommended nutrient control strategy. To implement the strategy, a detailed 3-D model has been developed and calibrated to evaluate various nutrient control scenarios of achieving the water quality goals set for the Chesapeake Bay.

TABLE 3
Eutrophication Control of Chesapeake Bay

Year	Activities
1976	Chesapeake Bay Program established a 7-year study
	Water quality problem identification
1983	7-year study report released
	Problems identified:
	Living resources
	Nutrients/eutrophication
	Toxics
	Submerged vegetation
1983	First Bay Agreement to clean up the Bay
	Implementation of study results
	Point source control of phosphorus
	Continuing monitoring and research
1987	Steady-state water model completed
	Second Bay agreement calling for 40% nutrient reduction (based on the model results) of point and nonpoint input by year 2000
	A new 3-D, time-variable water quality was launched to develop a nutrient control strategy
1988	Phosphate detergent ban complete in Bay watershed
	Phosphorus removal at POTWs
1992	3-D water quality model complete, ready for scenario evaluations
1990s	Tributary strategy
	Implementation

IV. ORGANIZATION AND SCOPE

The materials following this introduction are presented in seven chapters. Chapter 2 presents fundamental principals of estuarine water quality modeling. The major building blocks for an estuarine water quality model are mass transport, kinetics, and wasteloads. Only the first two components are discussed in this book. Chapter 3 then focuses on estuarine mass transport. BOD/DO and eutrophication model kinetics are presented in Chapter 4. Kinetics of toxicants are discussed in Chapter 5. Sediment-water interactions are important to many estuarine water quality problems and are presented in Chapter 6. Integration of estuarine hydrodynamic and water quality models is presented in Chapter 7. Lastly, mixing zone modeling is presented in Chapter 8.

Case studies, with intensive detail are presented throughout this book, demonstrating many successful model applications.

REFERENCES

1. **McCutcheon, S. C.,** *Water Quality Modeling, Volume I: Transport and Surface Exchange in Rivers,* CRC Press, Inc., Boca Raton, FL, 1989.
2. **Schroeder, S. H.,** Pollution control: changing approaches, *Environ. Sci. Technol.,* 15, 1287, 1981.
3. **Thomann, R. V.,** Principles of waste load allocations, in *Quality Models of Natural Water Systems,* Manhattan College Summer Institute in Water Pollution Control, 1973.
4. **U.S. EPA,** Guidance for Water Quality-based Decisions: The TMDL Process, Office of Water (WH-553), Washington, D.C., EPA/440/4-91-001, 1991.
5. **Pritchard, D. W.,** What is an estuary: physical viewpoint, in *Estuaries*, G. H. Lauff, Ed., American Association for the Advancement of Science, Publication No. 83, 2, 1967.
6. **Mills, W. B., Porcella, D. B., Ungs, M. J., Gherini, S. A., Summers, K. V., Mok, L., Rupp, G. L., and Bowie, G. L.,** Water Quality Assessment: A Screening Procedure for Toxic and Conventional Pollutants, Part II, EPA/600/6-85/002b, 142, 1985.

Chapter 2

FUNDAMENTALS OF ESTUARINE MODELING

I. PERSPECTIVE

Estuarine models can be classified into two groups: hydrodynamic models and water quality models. They are designed to calculate the concentration or distribution of a constituent, property of parameter in the estuary. To determine concentrations or distributions, transport processes and transformation processes must be resolved. Transport processes are basically hydrodynamic and include advection, turbulent diffusion, and, if spatial reduction is involved, dispersion. Transformation processes encompass the sources and sinks to which the parameter is subjected and may be physical, chemical, or biological. While this book addresses the water quality aspects, i.e., the kinetics or transformation, recognition is also given to hydrodynamics. However, in terms of modeling the hydrodynamics or water quality in estuaries, there is a fundamental difference, which leads to a profound impact on the advancement of the modeling technology for the hydrodynamic and water quality models.

First, consider the hydrodynamic models. It is generally accepted that estuarine hydrodynamics may be adequately described by the three momentum equations, the continuity equation, and the equation of state. These five equations have been the backbone of the hydrodynamic models in one-, two-, or three-dimensional configurations. Although parameterization is usually required to close the conservation of momentum equations, the momentum equations and continuity equation are derived from fundamental principles and have been widely accepted in the past two centuries.

On the other hand, the basis for water quality models is the conservation of mass equation. Kinetics formulations cannot be represented in a fashion similar to the momentum equations in hydrodynamic models. In fact, most kinetics are formulated using parameterization. As such, a high degree of empiricism is involved in kinetic formulation. Thus, water quality data in the prototype system is vital to the success of water quality modeling. Many kinetic formulations, and their coefficient values, need to be validated using the prototype data.

II. KEY COMPONENTS OF A WATER QUALITY MODEL

There are two principal components to a mathematical model.[1] One component is concerned with the specific water quality problem content — the particular water body (its geometry, flow, dispersion), and specific identification of the water quality problem. For example, when considering the estuarine eutrophication problem, specification of the relevant variables (e.g., chlorophyll *a*, phosphorus) must then be made and the kinetic linkage, or interactions between the variables, must be specified. This might include, for example, phosphorus uptake by phytoplankton and subsequent release upon phytoplankton death or nutrient release from the sediment for subsequent utilization by the phytoplankton. All these interactions and processes are formulated and the model parameters and kinetic coefficients then specified. Such specification for eutrophication, for example, might include numerical values for the phosphorus recycle rate, the nitrogen levels at which phytoplankton growth is inhibited and incoming solar radiation. Formulations of these interactions form a set of differential equations in time and space. The second major component of a model is a computational scheme or framework to calculate the concentrations of the state variables.

The equations describing the spatial and temporal distribution of water quality constituents are developed using the principle of mass conservation, including the inputs with the transport, transfer and reactions processes. The general expression for the mass balance equation about a specific volume, V is:

$$V\frac{dV}{dt} = J + \sum R + \sum T + \sum W \tag{2.1}$$

where

C = concentration of the constituent in V
J = mass transport through the system
R = reactions within the system
T = interphase transfers and
W = inputs

Topics such as mass transport, hydrodynamics, and kinetics are presented in the following chapters. Further, the water column kinetics are divided into BOD/DO, nutrient and eutrophication, and toxic substances. In addition, sediment-water interactions are described in a separate chapter.

III. MODEL DESIGN, CALIBRATION, VERIFICATION, AND POSTAUDIT

A. SPATIAL AND TEMPORAL SCALES

There are two aspects to the time and space scale determination:

1. The temporal and spatial extent of the water quality problem and variable
2. The temporal and spatial interval of the computation, i.e., the time step and spatial grid dimensions of the computational scheme

Figure 1 presents typical time and space scales for various water quality problems. The time scale of dissolved oxygen problems is in the order of days, which is understandable since the controlling reaction rates typically range from 0.05 to 1.0 d^{-1} (base e). Similarly, the spatial scale for dissolved oxygen is in the order of tens of miles.

Note that the computational time step may vary from minutes to days and the spatial grid may vary from hundreds of feet to miles for dissolved oxygen. In addition, the relevant spatial scale may also involve determination of the necessity to include vertical calculations in addition to horizontal calculations.[1] In general, short-term time scales (hours to days) are coupled with local and intermediate spatial scales (0.5 to 20 mi) and represent the principal responses due to highly transient random inputs such as combined sewer overflow and short-term nonpoint source inputs. The principal water quality variables include dissolved oxygen, coliform bacteria, and chlorine residual.

Time scales of month to month and year to year are generally coupled with regional to comprehensive spatial scales (75 to 100 mi) and represent the principal responses due to seasonally transient inputs such as nonpoint sources from the upper basin of the watershed and tributaries and point source inputs from treatment plants. The principal water quality variables include dissolved oxygen and the eutrophication/nutrient variables.

For more persistent materials such as toxic substances which decompose very slowly, much longer time scales (in decades) and spatial scales (entire basin) may be considered. Most toxic substances are accumulated in the sediment system which usually has much longer time and spatial scales than the water column.

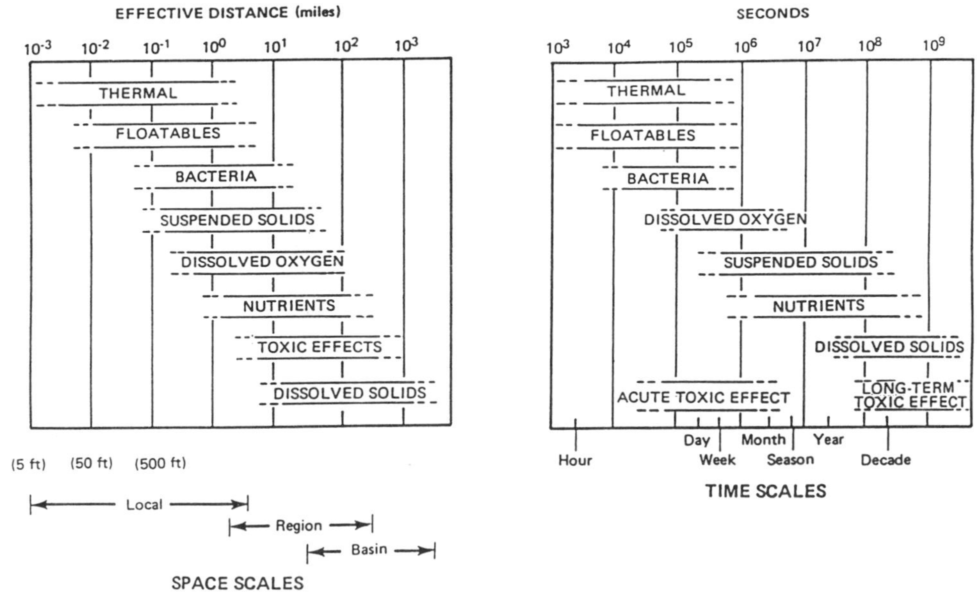

FIGURE 1. Spatial and temporal scales of water quality problems.

B. SPATIAL DISCRETIZATION

The box model (finite segment) approach is commonly used in spatial discretization. Several points should be considered in determining the length and configuration of model segments. In general, where receiving water concentrations showed or are expected to show a rapid rate of change of concentration, small segment lengths will be selected to reproduce these gradients. A second consideration in selecting segment sizes is numerical accuracy.

In box models, accuracy of the calculated solution is a function of segment size. The term associated with accuracy based on segment size is called numerical dispersion and can be viewed as an additional mixing or dispersion term in the longitudinal direction. To minimize errors, this term should be on the order of, or less than, the actual dispersion coefficients.

Segment sizes in model discretization should be appropriate to satisfy both criteria noted above. The segmentation network could be three-dimensional with surface, bottom and lateral segments in the water column.

C. MODEL CALIBRATION, VERIFICATION, AND SENSITIVITY ANALYSES

Model calibration is adjusting rate constants and coefficients to account for differences between laboratory and field conditions and/or site-to-site differences. This step requires field data for statistical comparison between field measurements and model calculations. It should be pointed out that in many cases, water quality models are planning tools, rather than predictive techniques. The model is used primarily to examine the spectrum of responses of the estuarine system which may occur under varying planning alternatives.[2] Thus, extreme accuracy is not required, but rather trends or incremental impact may be sufficient to provide answers to the water quality problems being addressed. It is thus the nature of the problem which primarily determines the degree of the calibration required. The degree of certainty required is determined by many factors, the most of which is the consequence of an erroneous decision.

Model verification is done by further model simulations using another field data set(s) collected under differing environmental conditions. Again, an acceptable statistical comparison between field measurements and model results should be achieved with the acceptance criterion established prior to model application. The relative complexity of the model is an important factor in its verification. That is, more complex models required a greater degree of verification. The complexity of the model is measured by the number of the transport, kinetic, and input terms in the equations.

D. MODEL POSTAUDIT

Model postaudit is a comparison of model forecasts with future field measurements as they are collected. Not many post-audit studies have been performed, but they are the ultimate test of the predictive nature of the model. In the case of BOD/DO modeling, the model formulations are usually empirical and a post-audit analysis is essential. For example, in quantifying the impact of reduced BOD loads from dischargers, analysts need to "guess" the deoxygenation rate in the receiving estuary as the wastewater characteristics, which is expected to change but is unknown prior to the load reduction. In this case, a postaudit is very important.

Another purpose of the postaudit is to evaluate the model performance following the treatment improvement of dischargers. For example, a typical audit of BOD/DO and eutrophication model addresses the following questions:

1. Do the actual dissolved oxygen (DO) data after the upgrade is installed generally reflect the basic principle of the model, i.e., does the DO go up when the BOD goes down?
2. To what degree is the model successful in predicting quantitatively the observed DO and other system variables in the model, such as phytoplankton biomass and nutrients?

IV. COMPUTING ENVIRONMENT

Today's microcomputer systems and work stations are capable of performing many water quality modeling tasks, including postprocessing of model results. In 1987, Lung[3] pointed out that using personal computers (PCs) for water quality modeling would be a promising option. Today, the advancement of computer hardware and software makes this option even more attractive. In fact, almost all the case studies of water quality modeling presented in this text, including some sophisticated eutrophication models, were run on microcomputers.

The number crunching speed of these computer systems is increasing rapidly, particularly from high performance numerical coprocessors. In the meantime, the FORTRAN compiler performance has excelled. For example, many excellent 32-bit compilers have been developed for microcomputers and have contributed to the proliferation of using microcomputers for water quality modeling. In fact, many well configured systems have the computational power matching that of mainframe systems.

REFERENCES

1. **Thomann, R. V.,** Overview of Potomac Estuary Modeling. Report submitted by HydroQual, Inc. to the Potomac Studies Technical Committee, Washington, D.C., 1980.
2. **O'Connor, D. J., Thomann, R. V., and Di Toro, D. M.,** Phytoplankton models and eutrophication problems, in *Ecological Modeling*, C.S. Russell, Ed., Resources for the Future, Inc., Washington, D.C., 153, 1975.
3. **Lung, W. S.,** Water quality modeling using personal computers, *J. Water Pollut. Contr. Fed.*, 59, 909, 1987.

Chapter 3

QUANTIFYING ESTUARINE MASS TRANSPORT

I. ESTUARINE HYDRODYNAMIC MODELING

In estuarine modeling, water movements (circulation) and mixing are required for water quality simulations. Accordingly, hydrodynamic computations are conducted to quantify the mass transport. However, it was considered sufficient to assume steady flow, complete mixing, and simple dilution in estuarine water quality modeling in the past. Attention was focused on the kinetics of reaeration, biodegradation, and the like, as in the classic case of the oxygen sag; hydrodynamics per se was not the concern of the analysts.[1] In the meantime, the computation of circulation in estuaries has been the subject of a large and continuing series of investigations. The motivation has been primarily hydrodynamic in nature; that is, the concern is directed at the circulation itself and the processes that generate and modify the velocities and stage. Two recent publications by Kjerfve[2,3] present a comprehensive treatment of estuarine hydrodynamics.

While conventional pollutants, such as BOD and nutrients, have been attracting much attention for the problems of depressed dissolved oxygen and accelerated eutrophication, toxic pollutants present in estuaries are becoming a major concern of the general public. Sediments play a major role in determining the fate and transport of toxic contaminants, many of which are quite persistent. As such, long term simulations of the water quality in an estuarine system are needed to address the loading-sediment flux relationships. Hence, there is an urgent need for reliable hydrodynamic solutions which in turn drive water quality models for these applications. The use of these models for other purposes, such as computation of the distribution of substances, of interest has more recently been attempted in water quality.

A review of hydrodynamic modeling is provided in this section to examine the available types of estuarine hydrodynamic models from three viewpoints: the hydrodynamic features of interest; the computational details, and *the water quality problems that can be addressed using these frameworks*. The sequence proceeds from the least spatially detailed to the most complex. No attempt is made to provide elaborate discussions on estuarine hydrodynamic modeling. Rather, the purpose is to present a perspective of hydrodynamic calculations and their relationships with water quality modeling.

A. ONE-DIMENSIONAL HYDRODYNAMIC MODELS

The appropriate equations of motion for a one-dimensional vertically and laterally averaged estuary are well known and have been applied to the calculation of the stage and discharge variation over a tidal cycle.[4] Irregular cross-sectional areas are routinely accommodated as are time-variable upstream inflows and downstream boundary conditions. Typically the nonlinear convective acceleration terms are neglected and the adjustable parameter is the possibly spatially varying bottom roughness coefficient. The object of the calculation is to compute the variation of stage and velocity over the tidal cycle. Stage variations are usually reproduced with great fidelity whereas the velocity variations are more difficult to reproduce.

Computationally the models are reasonably straightforward and the computational time required is not excessive. Both implicit and explicit methods have been employed with the former enjoying a computational speed advantage and unconditional stability for the linearized case.

The principal application to problems of water quality has been for the computation of salinity intrusion in estuaries. The model by Thatcher and Harleman[5] is a notable example of such a formulation and implementation. The significant additional parameter that must be included for the computation of mass transport is the dispersion coefficient, which is chosen to reproduce the salinity distributions.

B. TWO-DIMENSIONAL VERTICALLY AVERAGED HYDRODYNAMIC MODELS

The seaward portions of most estuaries have sufficiently large width so that a two-dimensional segmentation is more appropriate. Again, the tidal flow is the object of the calculation with stage and lateral, and horizontal velocities being computed. An early version of these models is the node-channel formulation applied to San Francisco Bay and Estuary by Shubinski et al.[6] using the Dynamic Estuary Model (DEM). More recent versions have been formulated using rectangular grids. These later models usually incorporate the convective acceleration terms and the Coriolis force in the momentum equation. Surface wind stress can usually be accommodated easily. The adjustable parameter is the bottom roughness coefficient. The stage variations can usually be reproduced quite well, whereas the velocity variations require extensive adjustment of the roughness coefficients.

The computational methods employed are again either implicit (actually alternating direction implicit) or explicit as in the node-channel formulation with the former enjoying a computational advantage in that the choice of integration time step is governed primarily by considerations and not by stability constraints. Time steps on the order or one to five minutes are typical.

These models have been successfully used for water quality investigations that can be conveniently divided into three categories:

- Intratidal calculations, for which the tidal flows are explicitly taken into account
- Intertidal calculations, for which only the net, nontidal flows are considered
- Steady-state calculations for which the stationary net nontidal flows are considered

For all these applications, in addition to the flow field, it is necessary to specify the dispersion coefficients, which are determined by analyzing either the salinity distribution or the motion of another tracer.

Intratidal calculations are appropriate for the detailed description of the distribution that results from highly time-variable inputs of mass such as a spill or a storm water overflow. On the other hand, intertidal calculations are more appropriate for the seasonal time scale that characterizes the intratidal hydrodynamic model calculations are averaged over the tidal cycle in order to produce the net nontidal flow distribution. The reason that this is possible is that the hydrodynamics are not coupled to the water quality parameters for these models. In particular, it is assumed that the salinity and temperature distributions do not affect the vertically averaged flow field, so that the hydrodynamic calculation can rapidly reach a periodic steady state without regard to the distribution of salinity and temperature. Typically, these net nontidal flow fields are computed every two weeks to one month and are incorporated into conventional water quality calculations. As with all such calculations, the dispersion coefficients are adjusted to reproduce the observed salinity.

For reasonably complex geometries these hydrodynamic calculations are very useful since it is uncertain just how the tidal action generates internal circulations and how the upstream freshwater flow is routed through the system. Nevertheless, since the computed flow field is vertically averaged, the mass transport mechanisms associated with the classical estuarine two-layer flow are not reproduced and this mixing is introduced via the assigned longitudinal and lateral dispersion coefficients. These are usually found to be flow dependent, as in the case presented above. Careful calibration and verification procedures are required in order that the appropriate dispersion coefficients are found and tested.

A good example of such models is an early Chesapeake Bay model developed by Blumberg.[7] A net circulation was obtained from the tidal computations. These indicated a net circulation induced by the tidal action. Extremely careful verification of the relevant circulation features are required if the results of these calculations are to be meaningful.

C. TWO-DIMENSIONAL LATERALLY AVERAGED DENSITY STRATIFIED HYDRODYNAMIC MODELS

Since the classical investigations of Pritchard[8] and Rattray and Hansen,[9] the vertical distribution of the horizontal velocity in an estuary has been the subject of a number of modeling studies. The usual approximation is to laterally average the equations of motion and retain the vertical component of the convective acceleration. The motion is driven by the pressure gradient variation due to the salinity stratification. Even a slight vertical stratification characteristic of partially mixed estuaries is sufficient to induce a net landward flow in the bottom waters and a net seaward flow in the surface. The hydrodynamic objectives of these calculations are usually to compute these velocities either intra- or intertidally together with the vertical velocity that results from continuity. The critical uncertainties are the vertical eddy viscosity of momentum, the vertical eddy diffusion of salt, and, if a bottom layer is also involved, the bottom shear stress.

Numerical models for such phenomena are usually intratidal in formulation.[7,10-15] A recent real-time model for the Patuxent Estuary was developed by Olson and Kincaid,[16] adopting a modeling framework by Wang and Kravitz.[17] The computational demands of the intratidal numerical schemes tend to be large, since the time to reach a realistic condition, either steady state if that is the condition being investigated, or the time to have reduced the effects of the necessarily imprecise initial condition specification, is on the order of one month since this is the time scale of longitudinal transport of salinity. On the other hand, the integration time step is necessarily small, on the order of one to two minutes, which is a result of the stability considerations of the finite difference forms of the hydrodynamic equations. Hence at best on the order of 10,000 integration steps are required to reach an acceptable solution. This is in contrast to the spin-up time of the two-dimensional vertically averaged models of on the order of one tidal cycle or 500 integration steps. The reason for this 200-fold difference is that vertically integrated hydrodynamic models are independent of the salinity whereas laterally integrated hydrodynamic models are directly dependent on the salinity and temperature fields which have inherently long characteristic time scales.

Laterally integrated intratidal models tend to produce the same type of horizontal velocity profiles as the intertidal models superimposed on the tidal oscillation which is also computed. Certain numerical difficulties due to abruptly terminating layers at the points of decreasing depth produce some numerical oscillations especially in the vertical velocities. In concept and physical content, however, these models are quite like the intertidal models.

A water quality problem that requires a specification of the vertical variation of the horizontal vertical velocity is the distribution of suspended solids in estuaries. The interaction of the particle settling velocity with the upward, landward, and the seaward velocities is responsible for the turbidity maximum observed at the tail of the salinity intrusion. The computed suspended solids, based on the velocities computed from the hydrodynamic model are compared to observation, if available. The detailed hydrodynamic velocity field is essential for such a calculation. An example of this type of calculation is presented by O'Connor and Lung[18] for suspended solids which is presented in a latter section of this chapter.

D. THREE-DIMENSIONAL HYDRODYNAMIC MODELS

The processes affecting estuarine circulation and mixing may be described using equations based on conservation of mass and momentum. The fundamental equations generally include the conservation of water mass (continuity), conservation of momentum, and conservation of constituent mass:

$$\frac{\partial u}{\partial x}+\frac{\partial v}{\partial y}+\frac{\partial w}{\partial z}=0 \tag{3.1}$$

$$\frac{\partial u}{\partial t}+\frac{\partial(uu)}{\partial x}+\frac{\partial(uv)}{\partial y}+\frac{\partial(uw)}{\partial z}=-\frac{1}{r}\frac{\partial p}{\partial x}+fv+\frac{\partial}{\partial x}\left[E_x\frac{\partial u}{\partial x}\right]+\frac{\partial}{\partial y}\left[E_x\frac{\partial u}{\partial x}\right]+\frac{\partial}{\partial z}\left[E_x\frac{\partial u}{\partial x}\right] \quad (3.2)$$

$$\frac{\partial u}{\partial t}+\frac{\partial(vu)}{\partial x}+\frac{\partial(vv)}{\partial y}+\frac{\partial(vw)}{\partial z}=-\frac{1}{r}\frac{\partial p}{\partial x}-fu+\frac{\partial}{\partial x}\left[E_x\frac{\partial v}{\partial x}\right]+\frac{\partial}{\partial y}\left[E_x\frac{\partial v}{\partial x}\right]+\frac{\partial}{\partial z}\left[E_x\frac{\partial v}{\partial x}\right] \quad (3.3)$$

$$\frac{\partial w}{\partial t}+\frac{\partial(wu)}{\partial x}+\frac{\partial(wv)}{\partial y}+\frac{\partial(ww)}{\partial z}=-\frac{1}{r}\frac{\partial p}{\partial z}+fw-g+\frac{\partial}{\partial x}\left[E_x\frac{\partial u}{\partial x}\right]+\frac{\partial}{\partial y}\left[E_x\frac{\partial u}{\partial x}\right]+\frac{\partial}{\partial z}\left[E_z\frac{\partial u}{\partial x}\right] \quad (3.4)$$

$$\frac{\partial C}{\partial t}+\frac{\partial(uC)}{\partial x}+\frac{\partial(vC)}{\partial y}+\frac{\partial(wC)}{\partial z}=\frac{\partial}{\partial x}\left[K_x\frac{\partial C}{\partial x}\right]+\frac{\partial}{\partial y}\left[K_y\frac{\partial C}{\partial y}\right]+\frac{\partial}{\partial z}\left[K_z\frac{\partial C}{\partial z}\right]+\Sigma S \quad (3.5)$$

where

t = time
p = pressure
g = gravitational acceleration
ρ = density
f = Coriolis frequency
E_x, E_y, E_z = turbulent diffusion coefficients for momentum
u, v, w = mean velocity components
x, y, z = rectangular coordinates
K_x, K_y, K_z = turbulent diffusion coefficients for mass
C = concentration of water quality constituent
S = constituent source/sink term

Currently available three-dimensional estuarine circulation models combine the features of the two types of two-dimensional models discussed above. Both lateral and vertical variations of the horizontal and vertical circulation are calculated at each longitudinal cross-section on an intratidal basis. Salinity and temperature are also calculated since they greatly influence the details of the vertical distribution of the velocity field. The parameters required for the calculation are identical to those for the above models and in principle the calculation is capable of describing the three-dimensional circulation and the distribution of temperature and salinity. It can be seen that the one-dimensional or two-dimensional hydrodynamic models are obtained by spatially averaging the three-dimensional equations.

Cheng and Smith[23] presented a survey of existing three-dimensional hydrodynamic and salinity transport models for estuaries and tidal bays. They also discussed the various methods to treat variations in the vertical and horizontal dimensions, and time dependency.

Recent efforts on three-dimensional hydrodynamic modeling of the Chesapeake Bay have been reported by Blumberg[19] and Johnson et al.[20] In Blumberg's work, the hydrodynamic results were seasonally averaged to drive a steady-state water quality model for the bay. The water quality model results were crucial to the decision of a 40% nutrient reduction from point and nonpoint sources around the bay region. Results from the work by Johnson et al. were used to drive the three-dimensional water quality model for the Chesapeake Bay in an effort to confirm the 40% nutrient reduction target and to assist future water quality management for the bay. (It should be pointed out that very few three-dimensional estuarine hydrodynamic models have been rigorously tested using field data.) In the study by Johnson et al., real-time calculations were extensively compared against the field data in terms of stage and velocity.

One of the difficulties in three-dimensional estuarine hydrodynamic modeling is the intensive computational effort. For example, the Chesapeake Bay hydrodynamic model is run on a super-computer. Limited access to such computation hardware prevents wide applications of these models to many water quality problems. Most recent development of three-dimensional model focuses on improving the computational efficiency. They include sigma coordinates, boundary fitting and turbulence closure of the hydrodynamic equations.[21,22] Turbulence closure modeling is useful for independently calculating the diffusion coefficient, particularly in the vertical direction. Some of these improvements are still in the research and development stage and have not been applied to prototype estuarine systems for water quality management use.

To this date, computational effort is still an important factor in estuarine hydrodynamic modeling, particularly three-dimensional calculations. As pointed out by Cheng and Smith,[23] for a fixed cost, the computing power increases five to ten times every five years. Two-dimensional models are used today as research, prognostic, and diagnostic tools for understanding estuarine processes, and for providing guidelines in management decisions. If the trend of increasing computing power continues in the next five to ten years, three-dimensional models can be and will be used as commonly as two-dimensional models are today.[2,3]

E. SUMMARY REMARKS ON ESTUARINE HYDRODYNAMIC MODELING

Hydrodynamic models are formulated with momentum equations, the continuity equation, and a salt balance equation. Early practices in hydrodynamic modeling required assigning or selecting eddy viscosity values for momentum exchange. (In one-dimensional hydrodynamic models which lump turbulence effects into a simple roughness coefficient, selecting a bottom friction coefficient is needed.) While these mixing coefficients cannot be precisely determined, initial estimates are needed to begin the calibration procedure. Some of the useful guidance available can be found in Bowie et al.[24] and Fischer et al.[25] McCutcheon et al.[26] reviewed the commonly used methods of computing vertical mixing.

While the major uncertainty of the model parameters is associated with the eddy diffusion coefficients for momentum and for salt, these equations have a profound theoretical basis. Bedford[27] reviewed the turbulence closure models for estuarine modeling. In recent years, two- or three-dimensional turbulence closure models have been employed in water quality problems.[20] These models are promising in the sense of universality, but are still in the stage of research and have not been fully tested.[26]

Finally, it is important to evaluate and determine the role of hydrodynamic modeling in an estuarine water quality study prior to conducting extensive hydrodynamic computations. In many cases, water column kinetics may be much more important than the physical transport in the system that the water quality response is not sensitive to the hydrodynamic processes. Thus, simplified mass transport calculations may be sufficient and appropriate to address the water quality problem context.

II. MASS TRANSPORT MODELING

As in the previous section on hydrodynamic modeling, this section presents mass transport models from the least spatially detailed to the most complex. Under each category of spatial resolution, examples of application are discussed. In addition, a comprehensive description of the box model approach is presented.

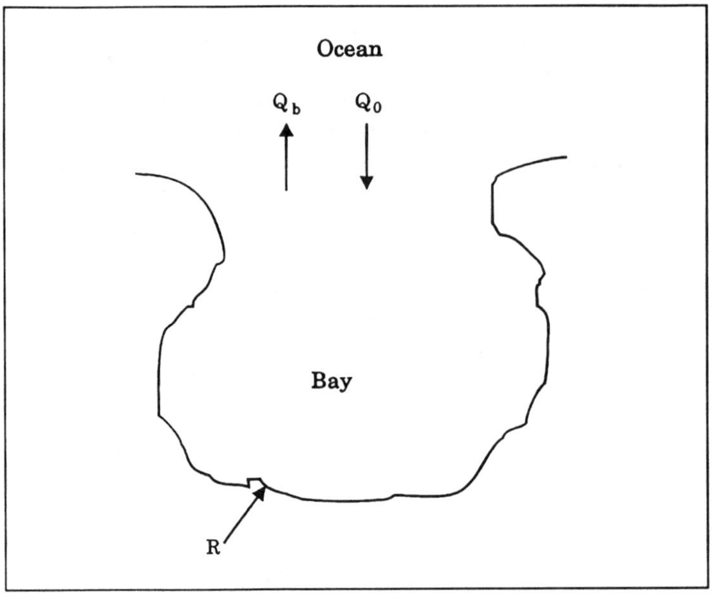

FIGURE 1. Zero-dimensional mass transport.

A. ZERO-DIMENSIONAL MASS TRANSPORT

The concept of a zero-dimensional model for flushing time calculation can be demonstrated using Figure 1. Let Q_i be the total volume of water coming into the embayment during flood tide. That is, Q_i is *new water* which has not been inside the embayment. Also, Q_o is the total volume of water going out of the embayment during ebb tide which does not return to the embayment. Under a tidally averaged steady-state condition, the salt balance can be expressed as:

$$Q_i S_i = Q_o S \tag{3.6}$$

where S_i is the salinity in the sea water and S is the average salinity in the embayment. Further, the water balance is

$$Q_i + R = Q_o \tag{3.7}$$

where R is the total freshwater inflow during the tidal cycle. Solving Equations 3.6 and 3.7 yields

$$Q_i = \frac{RS}{S_i - S} \tag{3.8}$$

$$Q_o = \frac{RS_i}{S_i - S} \tag{3.9}$$

Thus, both Q_o and Q_i may be determined from R, S_i and S, which can be readily measured.

Flushing rate can be defined as

$$r = \frac{Q_o}{V + P} \tag{3.10}$$

where

r = flushing rate, i.e., the fraction of water (or substance) in the embayment flushed out to the open sea per tidal cycle
P = tidal prism
V = volume at low tide

Tidal prism is the volume of water which enters the estuary or embayment during an incoming (flood) tide and equals high tide estuarine volume minus low tide volume and can be calculated by multiplying the tidal range by the surface area. Q_o can be related to P, tidal prism as

$$Q_o = \alpha P \tag{3.11}$$

Equation 3.11 indicates that the volume of water going out of the embayment is a fraction of the tidal prism. That is, a is between 0 and 1.0.

If the embayment receives large volume of freshwater to the extent that tidal prism becomes insignificant compared with the total volume of the embayment, then $V + P$ is approaching the tidally averaged volume. In that special case, Equation 3.10 becomes

$$r = \frac{Q_o}{V} \tag{3.12}$$

which is called *fraction of freshwater method* of calculating the flushing time.[28]

On the other hand, if the freshwater inflow is very insignificant, Equation 3.10 becomes

$$r = \frac{P}{P+V} \tag{3.13}$$

Equation 3.13 (with a = 1.0) is generally referred to as the *tidal prism method* of calculating the flushing time.[28]

Table 1 presents the results using these two methods to calculate flushing times in two embayments in Prince William Sound, Alaska.[29] In systems which have very small freshwater input (like Passage Cove and Snug Harbor in Table 1), the fraction of freshwater method yields very large flushing times (slow flushing rates). On the other hand, the tidal prism method calculates very small flushing times, considerably lower flushing times than calculation using the fraction of freshwater method. The exaggerated estimate of the rate of flushing is probably due to incomplete mixing of the estuarine water; the freshwater near the head of the estuary may not reach the mouth during the ebb.[28] Also, some of the water that does escape during the ebb could return on the following flood tide. Results in Table 1 indicate that a more rigorous method is needed to accurately calculate the flushing times for these embayments as presented in Section II. D.

B. ONE-DIMENSIONAL MASS TRANSPORT

The basic one-dimensional mass transport equation for steady state analysis can be derived as follows:[30-32]

$$\frac{1}{A}\frac{d}{dx}\left(EA\frac{dS}{dx}\right) - \frac{Q}{A}\frac{dS}{dx} - KS = 0 \tag{3.14}$$

where

S = water quality concentration
A = cross-sectional area
E = one-dimensional longitudinal dispersion coefficient
Q = net nontidal (freshwater) flow
K = first-order decay coefficient

TABLE 1
Flushing Time Calculation for Passage Cove and Snug Harbor

Parameter	Passage Cove	Snug Harbor
Tidally averaged volume (m^3)	4.017×10^6	143.421×10^6
Tidal range (m)	4.57	4.57
Surface area (m^2)	0.188×10^6	5.961×10^6
Tidal prism (m^3)	0.860×10^6	27.263×10^6
Low tide volume (m^3)	3.586×10^6	129.790×10^6
Freshwater flow (m^3/tidal cycle)	38379.6	127932.8
Open sea salinity, S_i (ppt)	32.3	32.3
Salinity in embayment, S (ppt)	30.0	30.0
Flushing time (tidal cycles[a]):		
Fraction of freshwater method	108.7	8974
Tidal prism method	1.72	5.76

[a] 1 tidal cycle = 12.54 h.

The longitudinal dispersion coefficient is a mixing coefficient that incorporates tidal flow oscillations, density effects, and other lateral and vertical velocity gradients (i.e., spatial reduction). In general, the tidal dispersion coefficient is determined with measured longitudinal salinity distributions. Many estuaries have been observed with an exponentially decreasing cross-sectional area, A, in the landward direction:

$$A = A_o e^{ax} \tag{3.15}$$

where A_o is the area at $x = x_o$ (downstream or ocean boundary) and a is a constant. A is approaching zero as $x = -\infty$. Substituting Equation 3.15 into Equation 3.14 for conservative substance ($K_d = 0$) such as salinity, C, yields:

$$E\frac{d^2C}{dx^2} + Ea\frac{dC}{dx} - U_o e^{-ax}\frac{dC}{dx} = 0 \tag{3.16}$$

where U_o = tidally averaged velocity at the downstream boundary (at $x = x_o$) and is equal to Q/A_o. Equation 3.16 can be solved with the following boundary conditions: $C = 0$ at $x = -\infty$ (upstream end) and $C = C_o$ at $x = x_o$ (downstream end). The solution is

$$C = C_o e^{\left[\frac{U_o}{aE}\left(e^{-ax_o} - e^{-ax}\right)\right]} \tag{3.17a}$$

and is derivation is presented in the appendix. For convenience, the ocean boundary (downstream end) is located at $x_o = 0$. Thus, Equation 3.17a becomes

$$C = C_o e^{\left[\frac{U_o}{aE}\left(1 - e^{-ax}\right)\right]} \tag{3.17b}$$

Figure 2 shows the derivation of the one-dimensional longitudinal dispersion coefficient, E, in the main channel of the Sacramento-San Joaquin Delta using Equations 3.15 and 3.17b and prototype cross-sectional area and specific conductivity data. In addition, the tidally averaged

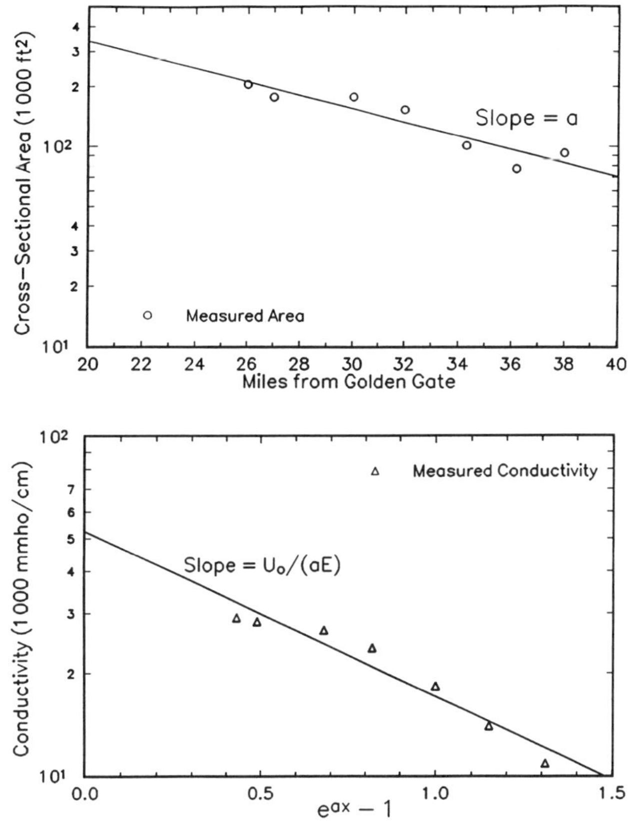

FIGURE 2. One-dimensional analytical solution for mass transport for Sacramento-San Joaquin Delta.

velocity at the downstream boundary is also available. First, the semilog plot of A versus x yields a best fit straight line with a slope equal to a = 0.04 mi.$^{-1}$ Next, the semilog plot of C (conductivity) versus $e^{-ax} - 1$ yields another best fit straightline with a slope equal to U_o/aE, which gives an E value of 15 mi^2/d.

For an infinitely long and constant parameter (e.g., Q, E, and A) estuary, Equation 3.14 can be simplified to:

$$E\frac{d^2S}{dx^2} - U\frac{dS}{dx} - KS = 0 \tag{3.18}$$

where $U = Q/A$, freshwater velocity which may be determined from the freshwater flow rate. The analytical solution of Equation 3.18 found in Thomann and Mueller[33] for salinity, C, (K = 0) is

$$C = C_o e^{-Ux/E} \tag{3.19}$$

It should be pointed out that the longitudinal dispersion coefficient in a tidally-averaged mass transport model is a quantity determined using salinity data collected under a given freshwater flow condition. This coefficient becomes less important as the accuracy of description of the advective transport increases.[34] For example, by incorporating vertical gradients of the longitudinal velocity, a two-dimensional mass transport analysis (see Section II.D.2) eliminates the one-dimensional longitudinal dispersion coefficient.

C. BOX MODEL FOR MASS TRANSPORT IN REAL ESTUARIES

In reality, very few estuaries are observed with constant (or even exponentially varying) cross-sectional area and constant flow, depth, and reaction rates. The preceding procedure is not applicable to derive the spatially variable dispersion coefficient. As such, to obtain analytical solutions of Equation 3.14 for many estuaries are difficult, if not impossible. Under such circumstances, the box or finite-segment model approach is employed. The box model was first used by Pritchard[35] for a two-layer estuarine system while the finite-segment approach was employed by Thomann[36] to model the Delaware Estuary under steady-state conditions. Both approaches are based on the same principle — using salinity distributions to determine the mass transport. The finite-segment approach has been widely used in water quality modeling practice for the past two decades. The box or finite-segment concept is rather simple and straightforward, particularly for the one-dimensional configuration. The procedure calls for dividing the estuary into a series of reaches, segments, or boxes, and writing the mass balance equation for each reach assuming the concentration gradient is not significant within the box. The approach is essentially a numerical finite difference approximation to the mass balance equation.[33] The advantages of the box model include:

1. The box/segment size could vary throughout the water body, according to the spatial resolution required to address the water quality problem.
2. The box/segment shape can be irregular, thereby could fit the physical boundary of a water body very well.
3. The box/segment model would include only ordinary differential equations as opposed to the partial differential equations to be solved by the finite difference approximation. Under steady-state conditions, the box model is reduced to a set of simultaneous algebraic equations.
4. Within its limitations the box/segment model can be applied to actual conditions in well-mixed and partially-mixed estuaries in which the horizontal net circulation exchanges are dominant over the horizontal nonadvective exchanges.[37]

The mass balance equation for a box model (Figure 3) is as follows:

$$V_k \frac{dC_{mk}}{dt} = \Sigma\left[-Q_{kj}\left(\alpha_{kj}C_{mk} + \beta_{kj}C_{mj}\right) + E'_{kj}\left(C_{mj} - C_{mk}\right)\right] - V_k K_{mnk} C_{mk} + W_{mk} \quad (3.20)$$

where

C_{mk} = concentration of water quality variable *m* in segment *k*, (M/L^3)
V_k = volume of segment *k*, (L^3)
Q_{kj} = advective flow from segment *k* to segment *j*, (L^3/T)
a_{kj} = finite difference weight, a function of the ratio of flow to dispersion, $0 < a < 1$,
b_{kj} = $1 - a_{kj}$
E'_{kj} = tidal mixing (exchange) flow between segments *k* and *j*, $(L^3/\text{sec}) = E_{kj}A_{kj}/\overline{L}_{kj}$
E_{kj} = dispersion coefficient between segments *k* and *j*, (L^2/sec)
A_{kj} = cross-sectional area between segments *k* and *j*, (L^2)
$\overline{L}_{kj}$ = average of characteristic lengths of segments *k* and *j*, $(L) = (L_k + L_j)/2$
L_k = characteristic length of segment *k* (L)
K_{mnk} = first-order reaction coefficient in segment *k* for interaction of water quality variable *m* with variable *n* $(1/T)$
W_{mk} = source (or sink) of variable *m* in segment *k*, (M/T) where *M, L, T* are the mass, length, and time units, respectively.

The first term on the righthand side of Equation 3.20 represents the mass entering or leaving (depending on the sign of the flow, *Q*) the segment *k* due to the nontidal flow. The second term

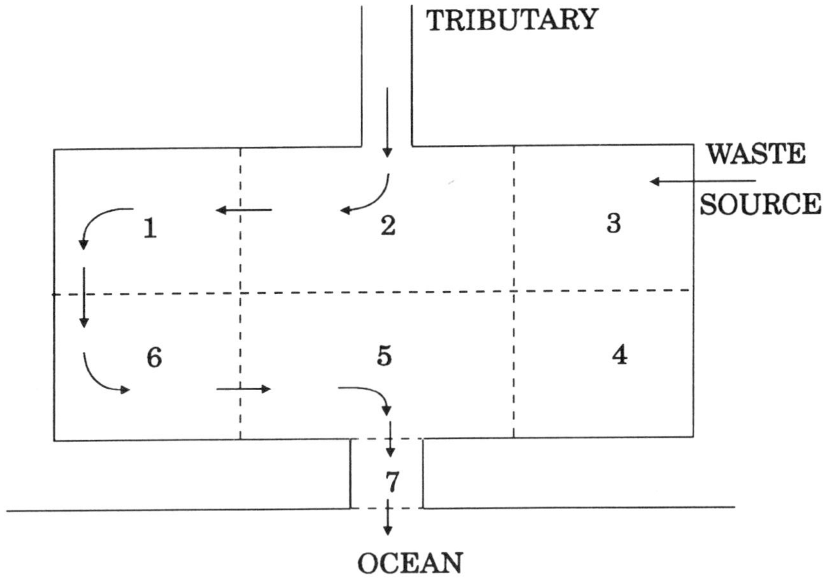

FIGURE 3. Box model configuration.

represents the tidal mixing effect principally due to the reversal of the tidal current. The sum of the flow and dispersion effect extends over all segments *j* bordering on segment *k*. The first-order decay (if any) of the variable is given by the third term, while the last term on the right-hand side of Equation 3.20 incorporates all direct sources and sinks of the variable, C_m.

If the estuary is now assumed to be at steady state, that is, all inputs, flow, exchange, and reaction rates are temporally constant from one tidal cycle to the next, then

$$V_k \frac{dC_{mk}}{dt} = 0$$

Next, consider a single variable whose concentration in segment *k* is C_k. If all terms involving the dependent variable, *C*, are grouped on the left-hand side, Equation 3.21 is obtained:

$$a_{kk}C_k + \Sigma a_{kj}C_j = W_k \qquad (3.21)$$

where

$$a_{kk} = \Sigma(Q_{kj}\alpha_{kj} + E'_{kj}) + V_kK_k$$

$$a_{kj} = Q_{kj}\beta_{kj} - E'_{kj}$$

For this discussion, it is assumed that there is only one kinetic reaction affecting this variable. K_k is the value of the kinetic rate in the k_{th} segment. Note that only the diagonal element contains this reaction term. A similar procedure is followed for each of the segments into which the water body has been divided. If *n* segments are used, then a series of *n* equations can be written.

Some special conditions will apply at the water boundaries of the system. For a section *k* where the flow between the boundary and the section is designated by Q_{kk} and is leaving the section (positive), and tidal mixing is designated by E'_{kk}, then a_{kk} becomes

$$a_{kk} = \Sigma(Q_{kj}\alpha_{kj} + E'_{kj}) + V_kK_k + Q_{kk}\alpha_{kk} + E'_{kk}$$

and the forcing (wasteloads) function is

$$W_k = W_k + (E'_{kk} - Q_{kk}\beta_{kk})\, C_b$$

where C_b is the boundary concentration of the variable C and is presumed known. In addition the source-sink term is augmented by the tidal mixing with the boundary.

For the flow entering the system from a boundary (Q_{kk} negative) the appropriate terms are

$$a_{kk} = \Sigma(Q_{kj}\alpha_{kj} + E'_{kj}) + V_k K_k + Q_{kk}\beta_{kk} + E'_{kk}$$

and

$$W_k = W_k + (E'_{kk} - Q_{kk}a_{kk})\, C_b$$

The n equations with suitable incorporation of boundary conditions (considered here as incorporated into the $W'_k s$) are given by

$$a_{11}C_1 + a_{12}C_2 + \ldots + a_{1n}C_n = W_1$$

$$a_{21}C_1 + a_{22}C_2 + \ldots + a_{2n}C_n = W_2$$

$$\cdots\cdots\cdots\cdots\cdots\cdots\cdots\cdots$$

$$a_{n1}C_1 + a_{n2}C_2 + \ldots + a_{nn}C_n = W_n$$

All a_{ij} and W_i quantities are assumed known in the above equation. The problem is to obtain the C_k values which represent the steady state spatial distribution of the water quality variable being considered. It is assumed that all segments interact with all other segments. This, of course, is not the case for actual systems for which there are generally six or less interfaces to each segment.

The theoretical development can be interpreted by writing the equations in a matrix form. Thus, one can express the n equations as

$$[A](C) = (W) \tag{3.22}$$

where $[A]$ is an $n \times n$ matrix and (C) and (W) are $n \times 1$ column vectors. The formal solution of Equation 3.22 is simply

$$(C) = [A]^{-1}\,(W) \tag{3.23}$$

Therefore, the problem of determining the spatial steady state water quality response in a multidimensional system reduces to solving a set of n simultaneous equations or determining the inverse of the system parameter matrix.

The above development is suitable for "single system" variables, i.e., water quality variables that are not forced by outputs from other quality systems. Examples of single system variables are salinity (where the reaction coefficient, K, is zero, representing a conservative variable), coliform bacteria, and biochemical oxygen demand (BOD), etc.

The procedure to quantify mass transport in a box model is an empirical one; e.g., the dispersion coefficient is determined by comparing a solution of the mass transfer equation with the measured concentration distribution of some substance in an estuary. The same dispersion coefficient is then used to predict the concentration distribution of some other substance. In practice, the empirical determination of dispersion coefficients is limited by the requirement that all source and sink (decay) terms for the substance must be known with reasonable accuracy. Therefore, this procedure is restricted to conservative substances such as salinity or to artificially introduced tracers for which decay and adsorption rates can be determined. In the past, observed data from either the prototype or a hydraulic model have been used to calibrate the dispersion coefficient. This methodology is relatively straightforward and widely used. A modeling framework, called HAR03, has been developed by Chapra and Nossa[38] to implement Equation 2.23 for multidimensional waterbodies under steady-state conditions.

It must be stressed that such a practice of back-calculating mass transport, either using Equation 3.17 or a box model, limits the model's credibility since a large degree of freedom can be used to adjust transport coefficients to fit the salinity data. As expected, the relative dependence on the advection and diffusion terms in the mass transport equation is a function of the scale of the model. As the scale increases, i.e., segments become larger or phenomena are averaged over longer time steps, the dependence on the empirical effective diffusion coefficient increases. Verification of coarse grid water quality models is usually accomplished by somewhat subjective adjustment of the coefficient until the model prediction agrees with prototype performance.[1] Nevertheless, this procedure is extremely useful in analyzing the mass transport in estuaries under existing conditions.

D. TWO-DIMENSIONAL MASS TRANSPORT

As discussed in hydrodynamic modeling, two-dimensional mass transport in many estuaries can be classified, in terms of the circulation pattern, into two categories: vertically homogeneous circulation or laterally homogeneous circulation. For both categories, intertidal results are more appropriate for the time scales that characterizes most water quality problems.

1. Two-Dimensional Vertically Integrated Mass Transport

For the two-dimensional vertically homogeneous calculations, the intratidal hydrodynamic model calculations are averaged over one or more tidal cycles in order to produce the net nontidal flow distribution. The reason that this is possible is that the hydrodynamics are not coupled to the water quality parameters for these models. In particular, it is assumed that the salinity and temperature distributions do not affect the vertically averaged flow field, so that the hydrodynamic calculation can reach a periodic steady-state rapidly without regard to the distribution of salinity and temperature. As with all such calculations, the dispersion coefficients are adjusted to reproduce the observed salinity. For reasonably complex geometries these hydrodynamic calculations are very useful since it is uncertain just how the tidal action generates internal circulations and how the upstream freshwater flow is routed through the system. Nevertheless, since the computed flow field is vertically averaged, the mass transport mechanisms associated with the classical estuarine two-layer flow are reproduced and this mixing is introduced via the assigned longitudinal and lateral dispersion coefficients.

The mass transport in the Sacramento-San Joaquin Delta is determined using this procedure.[39] First, the study area (Figure 4) is divided into 39 segments on a horizontal plane. The advective flows, Q_{kj}'s, were calculated from a hydrodynamic model. The next step is to determine the dispersion coefficients E'_{kj}'s in Equation 3.20 using data from a hydraulic model.

Salinity concentrations were measured in the main channel of the river and estuary portion after the model had reached equilibrium, i.e., the tidally averaged salinity in the system did not vary significantly from tidal cycle to tidal cycle. HAR03 was then used to calibrate the longitudinal coefficients in the main channel (see Figure 5).

Dye dispersion data from the hydraulic model was used to determine the lateral dispersion coefficients in the shallow bays: Grizzly Bay and Honker Bay and the dispersion coefficients between the shallow bays and the main channel. Following the system had reached steady state, the dye was released in the shallow bays and measured at a few locations at the high slack water for many cycles after the release. Thus, time-variable equation (Equation 3.20) was used to simulate the dynamic behavior of dye dispersion using constant advective flows and dispersion coefficients. The model results are compared with the dye data in Figure 5 for a freshwater outflow of 4400 *cfs*.

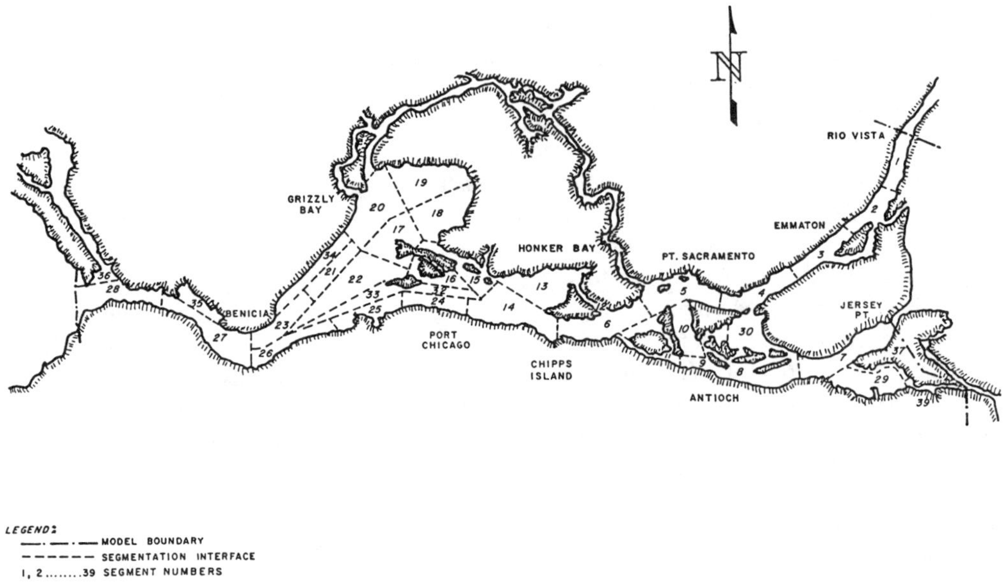

FIGURE 4. Two-dimensional segmentation for Sacramento-San Joaquin Delta.

2. Two-Dimensional Laterally Integrated Mass Transport

The two-dimensional laterally averaged mass transport equation under a tidally averaged condition is

$$\frac{\partial}{\partial x}(uS) + \frac{\partial}{\partial y}(vS) + \frac{\partial}{\partial x}(u_t S_t) = \frac{\partial}{\partial y}\left(\xi \frac{\partial S}{\partial y}\right) \tag{3.24}$$

where

$S =$ concentration of the water quality constituent
$u =$ velocity in the longitudinal direction
$v =$ velocity in the vertical direction
$u_t =$ tidal velocity along the longitudinal direction
$\xi =$ vertical dispersion coefficient on a tidally-averaged basis

Solving Equation 3.24 for the concentration, S, requires the quantification of the transport coefficients, u, v, and ξ.

The motion of the two-dimensional laterally averaged circulation is given by the pressure gradient variation due to the salinity stratification. Figure 6 displays the tidally averaged velocity and salinity profiles measured in a number of partially mixed estuaries: Sacramento-San Joaquin Delta, Rappahannock Estuary, James Estuary, and Patuxent Estuary. Pritchard[35] presented an analysis with the use of the box model approach for these partially stratified estuaries. In a close examination of Pritchard's work, Franco and Lung[40] indicated that the calculation of mass transport coefficients is quite sensitive to the local salinity levels and vertical gradients, particularly in the vicinity of the salinity tail.

Rattray and Hansen[9] independently developed solutions for this type of circulation and mixing, using analytical techniques. The hydrodynamic and mass transport equations are decoupled so that each may be solved separately and directly. Such analytical solutions have been given by Officer[41,42] and have been extended to include bottom frictional effects.

More recently, a more rigorous method of analysis was developed by Lung and O'Connor[43] and Lung.[44] Their methodology is based on the condition that the salinity distribution in both the longitudinal and vertical planes are known or may be assigned. Knowledge of the horizontal velocity permits determination of the vertical velocity from the continuity relationship.

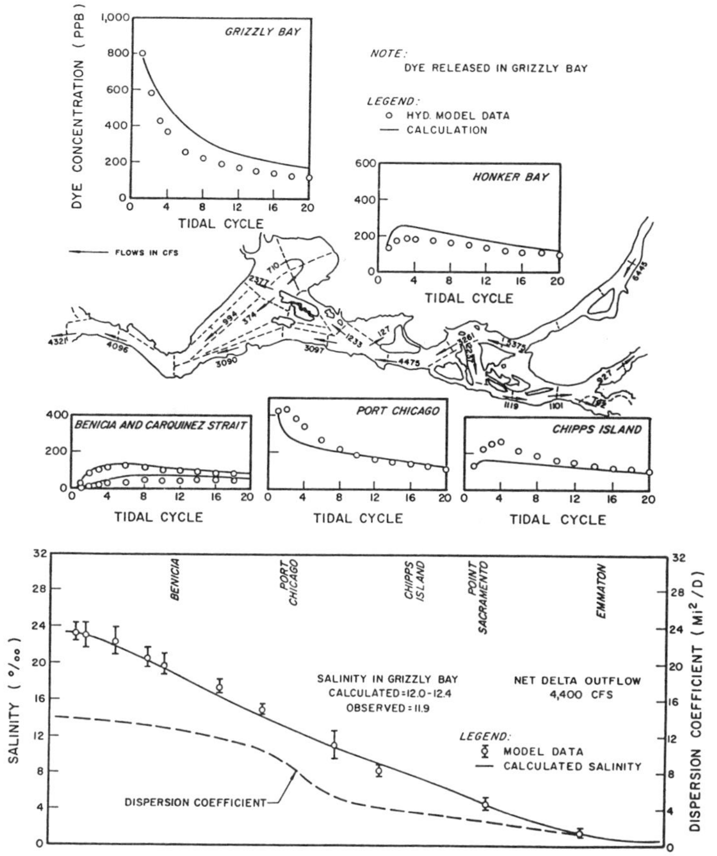

FIGURE 5. Two-dimensional mass transport analysis of Sacramento-San Joaquin Delta.

Under steady-state tidally averaged conditions, the longitudinal momentum equation for a laterally homogeneous estuary is

$$U_o \frac{\partial U_o}{\partial x} + \frac{1}{\rho}\frac{\partial p}{\partial x} = N \frac{\partial^2 u}{\partial y^2} \tag{3.25}$$

The vertical component of momentum equation is simply hydrostatic pressure equation:

$$\frac{1}{\rho}\frac{\partial p}{\partial y} = g \tag{3.26}$$

The equation of state is approximated as:

$$\rho = \rho_f (1 + \alpha C) \tag{3.27}$$

The equation of continuity can be expressed as:

$$\frac{\partial (bu)}{\partial x} + \frac{\partial (bv)}{\partial y} = 0 \tag{3.28}$$

The coordinate system is shown in Figure 7, in which the longitudinal x-axis is positive toward the ocean and the vertical y-axis is positive toward the bed of the channel. In these equations, U_o = amplitude of tidal current; r = density of the saline water; p = pressure; N = vertical eddy viscosity (assumed constant with depth); u = laterally averaged horizontal velocity; g = gravitational acceleration; r_f = density of freshwater; a = 0.000757 parts per thousand^{-1};

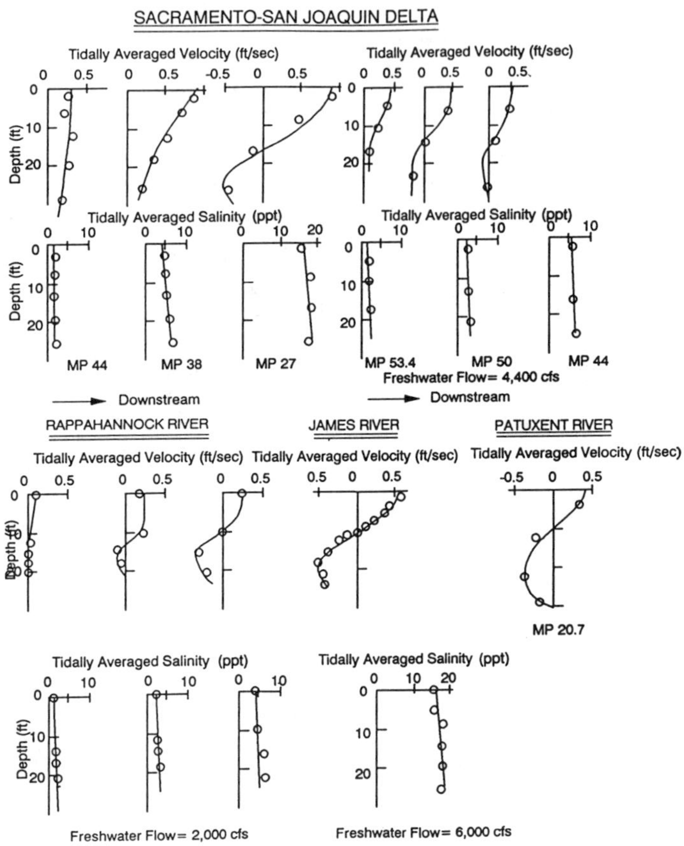

FIGURE 6. Tidally-averaged salinity and velocity profiles in various estuaries.

C = salinity (parts per thousand); b = depth-averaged width; and v = laterally averaged vertical velocity. Equation 3.25 may then be solved for the horizontal velocity, u, subject to the following boundary conditions: at the free surface ($y = -h$), $\partial u/\partial y = 0$, i.e., no wind effect; and at the bottom ($y = h$), $-N(\partial u/\partial y) = k|u_h|u_h$, in which k is a dimensionless friction coefficient = 0.0025; and u_h is the velocity at the bed.

Lung[44] introduced a linear approximation for the vertical variation of salinity:

$$C = C_s\,(1 + ay) \tag{3.29}$$

where C_s is the surface salinity and a is the vertical salinity gradient (ft^{-1} or m^{-1}).

Integrating Equation 3.26 yields

$$\int_{p_a}^{p} dp = g\int_{-\eta}^{y} \rho dy$$

where p_a is the atmospheric pressure at the free surface $y = -\eta$. Therefore,

$$p - p_a = g\int_{-\eta}^{y} \rho_f(1+\alpha C)dy$$

$$= g\int_{-\eta}^{y} \rho_f\left[1+\alpha C_s(1+ay)\right]dy$$

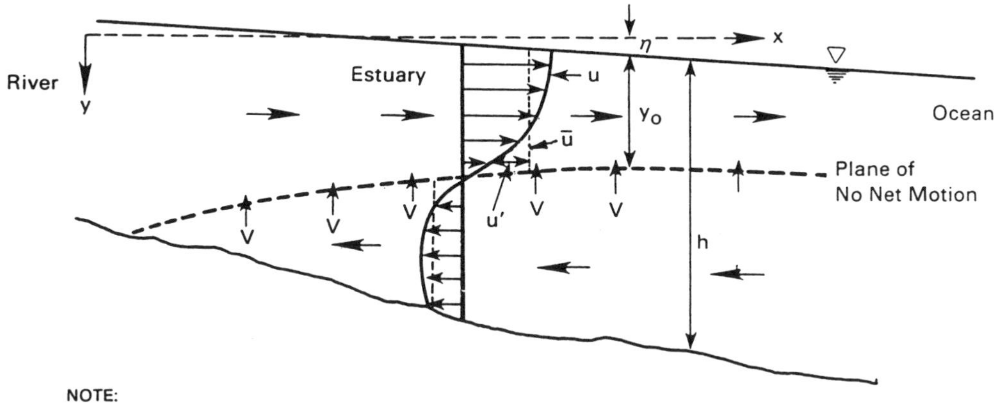

FIGURE 7. Coordinate system for two-dimensional estuarine circulation.

Further,

$$\frac{\partial p}{\partial x} = \rho_f g \int_{-\eta}^{y} (1+ay)\alpha \frac{dC_s}{dx} dy - \rho_f g\left[1+\alpha C_s(1-a\eta)\right]\frac{\partial(-\eta)}{\partial x}$$

$$= \rho_f g\alpha \frac{dC_s}{dx}\left[(y+\eta)+\frac{a}{2}(y^2-\eta^2)\right] + \rho_f g\left[1+\alpha C_s(1-a\eta)\right]\frac{\partial \eta}{\partial x}$$

and

$$\frac{1}{\rho}\frac{\partial p}{\partial x} = \frac{\rho_f g\alpha \frac{dC_s}{dx}\left(y+\frac{ay^2}{2}\right)}{\rho_f(1+\alpha C)} + \frac{\rho g\left[1+\alpha C_s(1-a\eta)\right]}{\rho_f(1+\alpha C)}\frac{\partial \eta}{\partial x}$$

$$= g\alpha \frac{dC_s}{dx}\left(y+\frac{ay^2}{2}\right) + g\frac{\partial \eta}{\partial x}$$

The above approximation is justified with the following assumptions:

1. α C is much smaller than 1
2. α C_s is much smaller than 1
3. a η is much smaller than 1

It should be pointed out that *a*, the vertical salinity gradient (ft^{-1}), does not vary with *x* significantly in the saline zone such that *da*/*dx* is small. This assumption may not be valid in the tail end of salinity intrusion.

Equation 3.25 now becomes

$$N\frac{\partial^2 u}{\partial y^2} = g\alpha \frac{dC_s}{dx}\left(y+\frac{ay^2}{2}\right) + g\frac{\partial \eta}{\partial x} + U_o\frac{\partial U_o}{\partial x} \quad (3.30)$$

Dividing Equation 3.30 through by N yields

$$\frac{\partial^2 u}{\partial y^2} = \frac{g\alpha}{N}\frac{dC_s}{dx}\left(y + \frac{ay^2}{2}\right) + \frac{g}{N}\frac{\partial\eta}{\partial x} + \frac{U_o}{N}\frac{\partial U_o}{\partial x} \tag{3.31}$$

Integrating Equation 3.31 gives

$$\frac{\partial u}{\partial y} = \frac{g\alpha}{N}\frac{dC_s}{dx}\left(\frac{y^2}{2} + \frac{ay^3}{6}\right) + \left(\frac{g}{N}\frac{\partial\eta}{\partial x} + \frac{U_o}{N}\frac{\partial U_o}{\partial x}\right)y + C_1 \tag{3.32}$$

The integration constant $C_1 = 0$ since $\partial u/\partial y = 0$ at $y = -h = 0$. Integrating again,

$$u = \frac{g\alpha}{N}\frac{dC_s}{dx}\left(\frac{y^3}{6} + \frac{ay^4}{24}\right) + \left(\frac{g}{N}\frac{d\eta}{dx} + \frac{U_o}{N}\frac{\partial U_o}{dx}\right)\frac{y^2}{2} + C_2 \tag{3.33}$$

C_2 is an integration constant which can be derived using the following condition:

$$Q/b = \int_h^{-\eta} u\, dy$$

where Q is the freshwater flow rate and b is the average width.

The solution of Equation 3.25 derived by Lung[44] is

$$u = \frac{g\alpha}{N}\frac{dC_s}{dx}\left(\frac{y^3}{6} + \frac{ay^4}{24} - \frac{h^3}{24} - \frac{ah^4}{120}\right) + \frac{g}{N}\left(\frac{d\eta}{dx} + \frac{U_o}{g}\frac{\partial U_o}{\partial x}\right)\left(\frac{y^2}{2} - \frac{h^2}{6}\right) + \frac{Q}{bh} \tag{3.34}$$

The water surface gradient, $d\text{h}/dx$ in Equation 3.34 needs to be evaluated prior to calculating u. Or, let $\text{t} = d\text{h}/dx + (U_o/g)\,(\partial U_o/\partial x)$. t can be evaluated by applying the remaining boundary condition at the channel bed: $N(\partial u/\partial y) = -k|u_h|u_h$, in which u_h is the bottom velocity at $y = -\text{h}$ and k is the bottom friction coefficient (= 0.0025). A bottom velocity in the landward direction ($u_h < 0$) may be obtained from Equation 3.34 as follows:

$$|u_h| = -\left[\frac{g\alpha}{N}\frac{dC_s}{dx}h^4\left(\frac{1}{8h} + \frac{a}{30}\right) + \frac{gh^2\tau}{3N} + \frac{Q}{bh}\right] \tag{3.35}$$

Subsequently

$$k|u_h|u_h = -k\left[\frac{g\alpha}{N}\frac{dC_s}{dx}h^4\left(\frac{1}{8h} + \frac{a}{30}\right) + \frac{gh^2\tau}{3N} + \frac{Q}{bh}\right]^2 \tag{3.36}$$

$N\,(\partial u/\partial y)$ at $y = -\eta$ is derived from Equation 3.34 as

$$N\left(\frac{\partial u}{\partial y}\right)_{y=h} = g\alpha\frac{dC_s}{dx}\left(\frac{h^2}{2} + \frac{ah^3}{6}\right) + gh\tau \tag{3.37}$$

Incorporating Equations 3.36 and 3.37 into the boundary condition yields

$$k\left[\frac{g\alpha}{N}\frac{dC_s}{dx}h^4\left(\frac{1}{8h} + \frac{a}{30}\right) + \frac{gh^2}{3N}\tau + \frac{Q}{bh}\right]^2 = gh\left[\alpha\frac{dC_s}{dx}\left(\frac{h}{2} + \frac{ah^2}{6}\right) + \tau\right] \tag{3.38}$$

Equation 3.38 is a quadratic equation of τ. It, therefore, has two roots, one of which can be discarded because it yields a seaward bottom velocity. The appropriate root is then substituted into Equation 3.34 to calculate the horizontal velocity as a function of depth. The righthand side of Equation 3.38 must be greater than zero, implying

$$\alpha\frac{dC_s}{dx}\left(\frac{h}{2} + \frac{ah^2}{6}\right) + \tau > 0 \tag{3.39}$$

In a separate study, Fischer[45] derived the same equation as a criterion for a landward bottom velocity. To use Equation 3.34, one needs to assign salinity gradients and the associated freshwater flow, channel characteristics at *selected* locations throughout the saline zone. Next, the surface salinity, C_s, and the longitudinal salinity gradient, dC_s/dx are derived from tidally averaged salinity data. The aforementioned parameters and a first estimate of eddy viscosity, N, are substituted into Equation 3.34 to obtain the vertical profile of horizontal velocity at each station. Within the range of reported values for estuaries, the eddy viscosity is adjusted such that the calculated vertical profile of horizontal velocity agrees with the data at a particular station. Examples of applying Equation 3.34 to a number of estuaries ranging from a laboratory flume to the James Estuary are presented in Figure 8. Basically, measured velocity data were used to determine the vertical eddy viscosity, N.

3. Two-Layer Box Model Approximation of Mass Transport

Equation 3.34 also yields the depth at which the net horizontal velocity is zero, the point of no net motion. The plane of no net motion is defined for the entire saline zone by interpolation forming a two-layer system. The vertically averaged velocity in each layer, u, is obtained from Equation 3.34; thus

$$u = \frac{1}{y_o + \eta} \int_{-\eta}^{y_o} u \, dy \tag{3.40}$$

in which y_o is depth of zero velocity. The estuary is further segmented horizontally and the horizontal flow in the surface and bottom layers at each vertical cross-section is calculated. The vertical velocity at y_o may be evaluated by means of Equation 3.28; thus

$$v(x, y_o) = -\frac{1}{b}\frac{\partial}{\partial x}\left(b \int_{-\eta}^{y_o} u dy \right) = -\frac{1}{b}\frac{\partial}{\partial x}[b(y_o + \eta)u] \tag{3.41}$$

Equation 3.41 indicates that the difference in horizontal flow between two adjacent vertical planes gives the vertical flow between the surface and bottom layer. The average vertical velocity is obtained by dividing the vertical flow by the average horizontal interfacial area of the segment.

An initial estimate of the vertical dispersion coefficient may be obtained from the empirical relationship with the eddy viscosity, N; thus

$$\xi = \frac{N}{1 + R_i} \tag{3.42}$$

in which R_i is the Richardson number defined as

$$R_i = g \frac{\partial \rho / \partial y}{\rho \left[\frac{\partial u}{\partial y} \right]^2} \tag{3.43}$$

It should be pointed out that in many partially mixed estuaries, the vertical variation of the horizontal velocity and the vertical dispersion are employed to characterize the transport pattern in lieu of the longitudinal dispersion coefficient in a one-dimensional configuration. In other words, the back-calculation of the one-dimensional dispersion coefficient using salinity distribution is no longer needed in a two-dimensional analysis. Thus, the two-dimensional mass transport coefficients can be readily derived from salinity distribution using a much more robust approach.

The next step of computation is to confirm these mass transport coefficients. Naturally, the derived coefficients are used to generate the salinity distributions in a two-layer fashion for comparison with the known salinity distributions. By means of the layer averaging process, Equation 3.24 can be approximated by a two-layer mass transport model for salinity as[43]

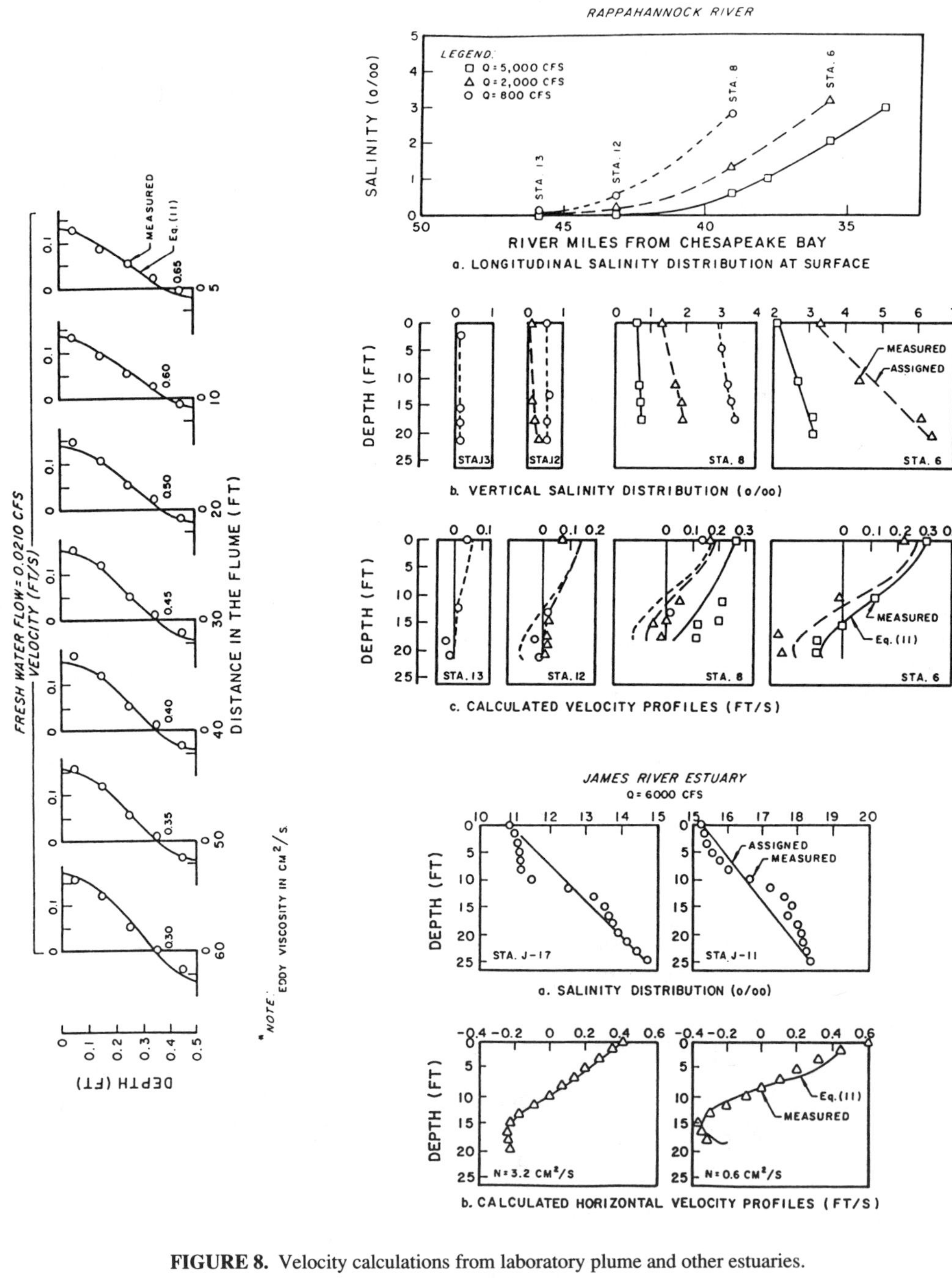

FIGURE 8. Velocity calculations from laboratory plume and other estuaries.

$$\frac{d}{dx}(u_1C_1)+\frac{d}{dx}\left(E_x\frac{dC_1}{dx}\right)+\frac{v(x,y_o)}{l_y}(\beta C_1+\gamma C_2)-\frac{\xi}{l_y^2}(C_1-C_2)=0 \tag{3.44a}$$

$$\frac{d}{dx}(u_2C_2)+\frac{d}{dx}\left(E_x\frac{dC_2}{dx}\right)-\frac{v(x,y_o)}{l_y}(\beta C_1+\gamma C_2)+\frac{\xi}{l_y^2}(C_1-C_2)=0 \tag{3.44b}$$

in which the subscripts 1 and 2 refer to the quantities in the surface layer and bottom layer, respectively; l_y is the average depth of two layers; and β and γ are weighting factors to

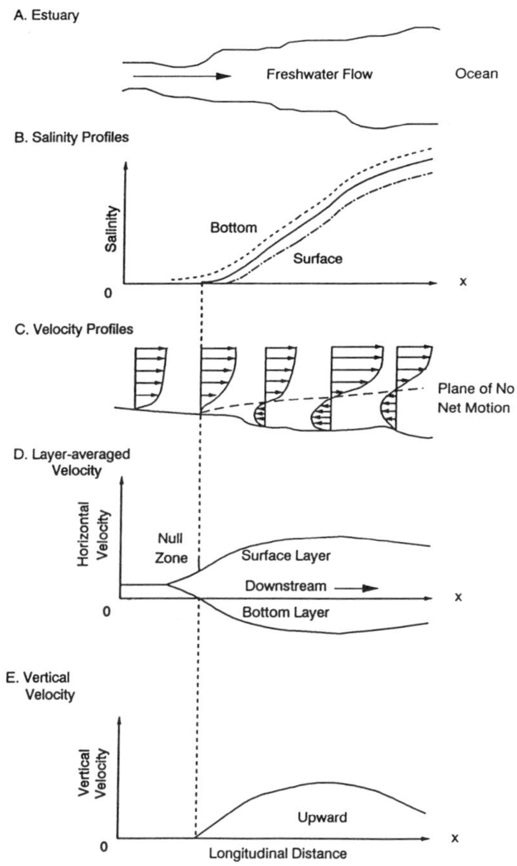

FIGURE 9. Schematic diagram and two-layer mass transport calculation procedure.

approximate the concentrations at the surface layer-bottom layer interface. E_x is a dispersion coefficient resulting from the layer-averaging process.[43] Equations 3.44a and 3.44b are the working equations of this mass transport framework. A typical schematic diagram of the two-layer box model configuration is shown in Figure 9. The plane of no net motion is usually used as the dividing line between the two layers. Successful applications of this methodology have been reported for the following estuarine systems: the Sacramento-San Joaquin Delta,[39] the Patuxent Estuary,[46] and the Hudson Estuary.[47] An independent evaluation of this methodology[48] reported that the analytical solution gives good results for the St. Lawrence Estuary in Canada.

The calculation is, in principle, simple and straightforward, and is briefly summarized in the following steps:

1. The assignment of salinity gradients and the associated freshwater flow, channel characteristics at selected locations throughout the saline zone, is first made. The surface salinity, C_s, and the longitudinal salinity gradient, dC_s/dx are calculated from tidally-averaged salinity data.
2. The aforementioned parameters and a first estimate of eddy viscosity, N, are substituted into Equation 3.34 to obtain the vertical profile of horizontal velocity at each station. Within the range of reported values for estuaries, the eddy viscosity is adjusted such that the calculated vertical profile of horizontal velocity agrees with the data from a particular station.
3. The point of no net motion is known from the aforementioned calculation at each station where a velocity calculation is made. The plane of no net motion is defined for the entire saline zone to form a two-layer system.

4. The average horizontal velocity in each layer is then determined from the velocity profiles calculated in Step 2. The estuary is further segmented horizontally and the horizontal flow in the surface and bottom layers at each vertical cross section is calculated.
5. The difference in horizontal flow between two adjacent vertical planes gives the vertical flow between the surface and bottom layer. The average vertical velocity is obtained by dividing the vertical flow by the average horizontal interfacial area of the segment. The procedure is a solution of the hydraulic continuity of Equation 3.28.
6. The empirical relationship in Equation 3.42 is used to provide an initial estimate of vertical dispersion at each station.
7. The advective transport derived in Steps 4 and 5 is incorporated with the vertical dispersion in Equation 3.44. The average salinity in both layers is calculated for comparison with the data. The aforementioned procedure, including velocity and salinity calculations, may be iterated in order to reproduce the observed salinity distribution and thus, to obtain the appropriate transport pattern.

4. Mixing Analysis of Embayments

Results presented in Table 1 indicate that the zero-dimensional calculations are not suitable for the mixing analysis of Passage Cove and Snug Harbor. A close examination of temperature and salinity measurements suggested that vertical gradients of temperature and salinity exist in these two embayments. Thus, the methodology developed by Lung and O'Connor[43] was used to calculate the two-layer mass transport in Passage Cove and Snug Harbor. Figure 10 shows the four-segment configuration used for these two systems with mass transport coefficients calibrated. The calculated and measured salinity levels in the segments are summarized in Table 2. The calculated residence times in the segments are also shown in Table 2. The total residence time is slightly longer than the flushing time generated by the tidal prism method but substantially less than the flushing time caused solely by freshwater dilution (see Table 1).

5. Modeling Turbidity Maxima

The temporal and spatial distribution of suspended solids has received increased attention of engineers and scientists in many parts of the world over the past few decades.[49-51] Suspended solids affect the transmission of light and thus, the growth of phytoplankton and other plants. They provide sites for the growth of microorganisms which impact water quality. They absorb heavy metals and pesticides and thereby influence the concentration of these substances both in the bed and in suspension. In estuaries, suspended solids are particularly significant partly because of the cohesiveness of the solids in saline waters and the characteristic circulation pattern which increases the retention of solids in these systems. In addition to the velocity imparted by the typical two-dimensional estuarine circulation, suspended particles possess a vertical settling velocity. If the vertical water velocity is downward it enchances the vertical flux of solids in that direction; by contrast, if the water velocity is upward, it tends to cancel the settling velocity. Thus, for the larger and more dense particles, such as sand and silts, the net vertical velocity is in the downward direction, but less than the settling velocity and for the smaller and less dense particles, such as clays and organics, the net velocity is directed upward, if the water velocity is greater than the settling velocity. Such vertical movement, in conjunction with the convergence of the landward-flowing density current and the seaward-flowing surface river current at the tail of salinity intrusion, is responsible for the solids concentrations in the saline zone of the estuary, which are greater than those of the upstream freshwater inflow and downstream density current. This phenomenon, referred to as the *turbidity maximum*, decreases with decreasing river flow and is frequently washed out under relatively high flow conditions.

The waters of the study area are known as the Western Delta-Suisan Bay from the interface between the saline water of San Pablo Bay at the northern end of San Francisco Bay, and the

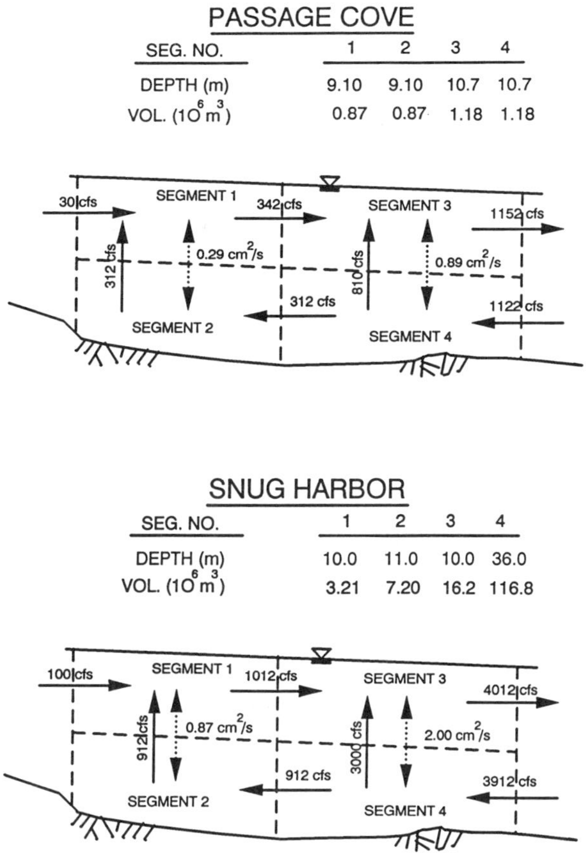

FIGURE 10. Segmentation and mass transport for Passage Cove and Snug Harbor.

TABLE 2
Mass Transport Calculation for Passage Cove and Snug Harbor

Model segment	Calculated salinity (ppt)	Measured salinity (ppt)	Calculated residence time (tidal cycles)
Passage Cove:			
Segment 1	29.47	28.75 (26.01–30.76)[a]	1.93
Segment 2	32.29	32.01 (30.80–32.33)[a]	2.12
Segment 3	31.46	29.79 (26.00–30.45)[a]	0.78
Segment 4	32.30	32.12 (30.80–32.40)[a]	0.80
		Total:	5.63
Snug Harbor:			
Segment 1	29.07	29.75 (26.75–31.33)[a]	2.26
Segment 2	32.20	32.04 (32.00–32.32)[a]	5.57
Segment 3	31.50		2.45
Segment 4	32.26		18.07
		Total:	28.35

[a] Range.

freshwater outflow from California's Central Valley, drained by the Sacramento and San Joaquin Rivers. Most of the freshwater flows enter the area from the inland drainage basin through the Sacramento and San Joaquin Rivers. High flows occur typically during the months of January through March because of seasonal rainfall and snowmelt patterns and drop to lower volumes during the remainder of the year. The water temperature in the study area is relatively uniform in both longitudinal and vertical directions.

To accommodate particle settling, Equations 3.44a and 3.44b are modified as follows:

$$\frac{d}{dx}(u_1C_1)+\frac{d}{dx}\left(E_x\frac{dC_1}{dx}\right)+\frac{v(x,y_o)+v_s}{l_y}(\beta C_1+\gamma C_2)-\frac{\xi}{l_y^2}(C_1-C_2)=0 \qquad (3.45a)$$

$$\frac{d}{dx}(u_2C_2)+\frac{d}{dx}\left(E_x\frac{dC_2}{dx}\right)-\frac{v(x,y_o)-v_s}{l_y}(\beta C_1+\gamma C_2)+\frac{\xi}{l_y^2}(C_1-C_2)=0 \qquad (3.45b)$$

where v_s is the net particle settling velocity in the water column on a tidally averaged basis and C_1 and C_2 are suspended solids concentrations in the top and bottom layers, respectively. The boundary conditions along the longitudinal axis are the concentrations associated with the freshwater and density input. The vertical boundary conditions are established by the bed-water interaction. A net flux from the water column to the bed implies that settling is greater than the scour and conversely a net flux from the bed to the water stipulate a scour greater than settling. In either case, a steady state may be achieved in which the rates are equal and the net flux is zero on a tidally averaged basis. This is the bed condition adopted in the analysis.

Figure 11 shows salinity and suspended solids results from the model applications to the Sacramento-San Joaquin Delta for freshwater flows of 4400 and 10,000 *cfs*. The average settling velocity used in the analysis is 8 ft/d on a tidally-averaged basis.[18] The model reproduces the turbidity maxima for these two flow conditions reasonably well. One should note that the two-layer mass transport is more pronounced with the higher flow than the lower flow. On the other hand, salinity intruded further upstream under a lower flow, which tend to increase mixing in the vertical direction, resulting in smaller salinity gradients in the water column.

6. Two-Layer Mass Transport in Patuxent Estuary

The Patuxent River has been a major focal point for water quality management in Maryland for over a decade. To assist the water quality management, a water quality model was recently developed to evaluate nutrient controls.[52] As part of the development of the water quality model (see Chapter 4), two-layer mass transport calculations were conducted to assist the calibration of water column kinetics. An important feature in the calculations was to track the tidally averaged two-layer mass transport on a time-variable basis for a period of three years (1983 to 1985). An extensive data base was used to develop the transport patterns. Basically, 14 to 19 slack water salinity surveys of the estuary in each year provided two-dimensional salinity distributions. Data from each survey was utilized to generate a two-layer mass transport pattern in a 38-segment configuration (Figure 12). These patterns throughout the three-year period were then incorporated into a series of time-dependent patterns as preliminary estimates of the transport coefficients. Subsequent adjustments of the transport coefficients in the two-layer structure reproduce the observed salinity distributions. Figure 13 shows the temporal plots throughout the three-year period for eight locations in the estuary. It is seen that the seasonal distributions of salinity in the Patuxent Estuary are closely reproduced by the two-layer mass transport model. The temporal variations of salinity in the estuary vary with the freshwater flow at the head of the estuary. High freshwater flows in the spring months reduce the salinity concentrations in the estuary while generating appreciable vertical salinity gradients. Somewhat reduced freshwater flows in the summer progressively raise the salinity levels. A select spatial

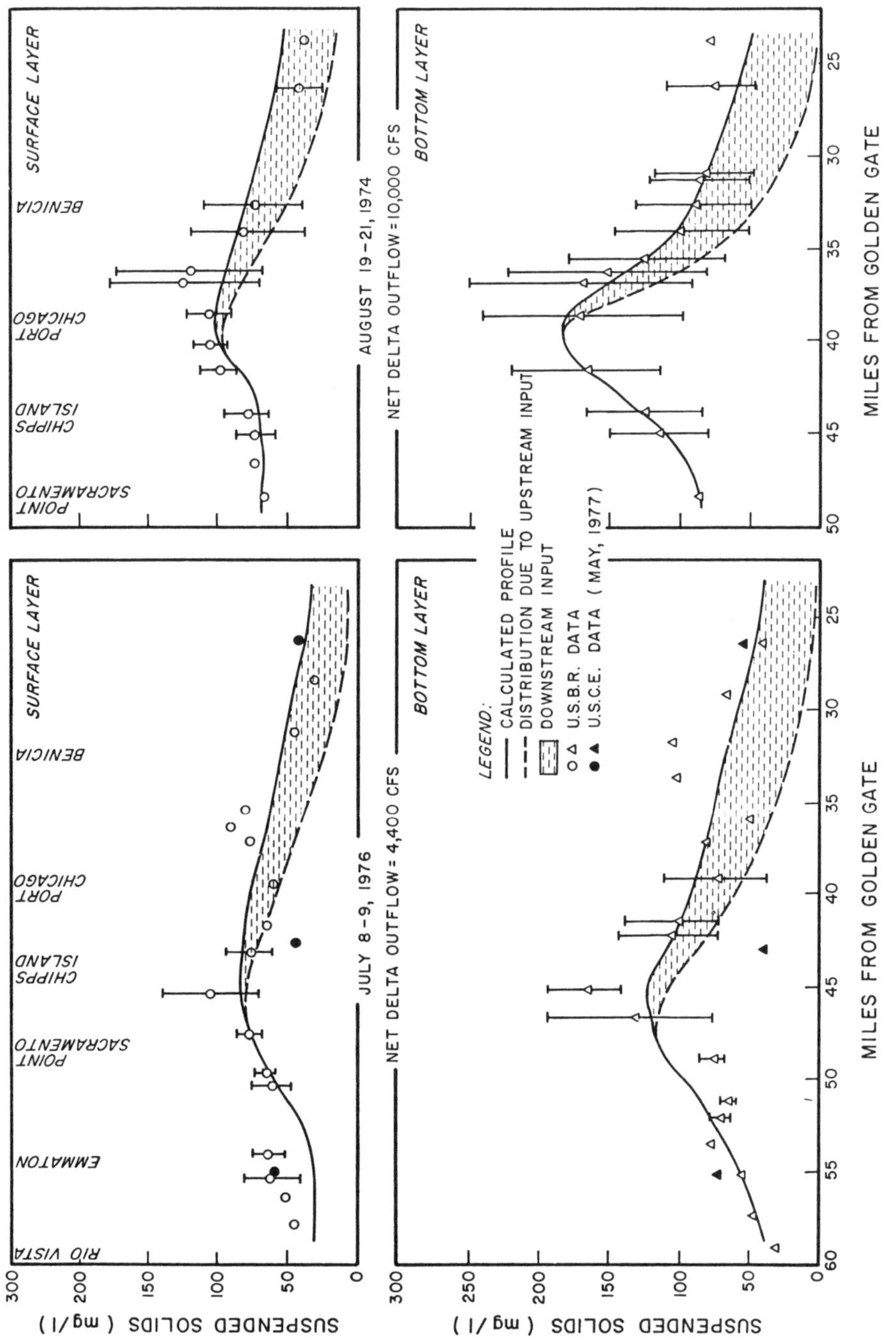

FIGURE 11. Two-dimensional suspended solids analysis of Sacramento-San Joaquin Delta.

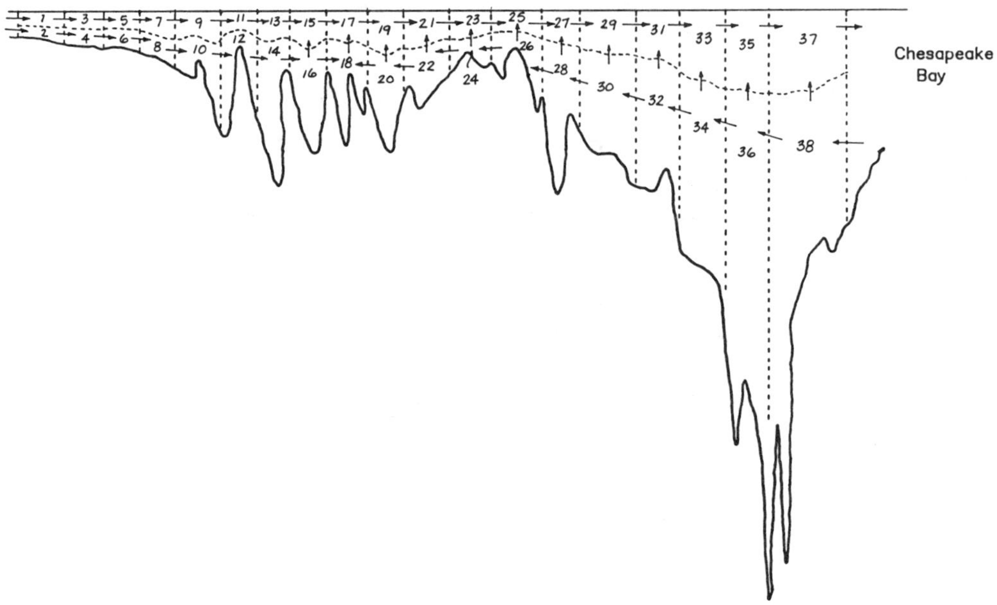

FIGURE 12. Patuxent Estuary segmentation and bottom bathymetry.

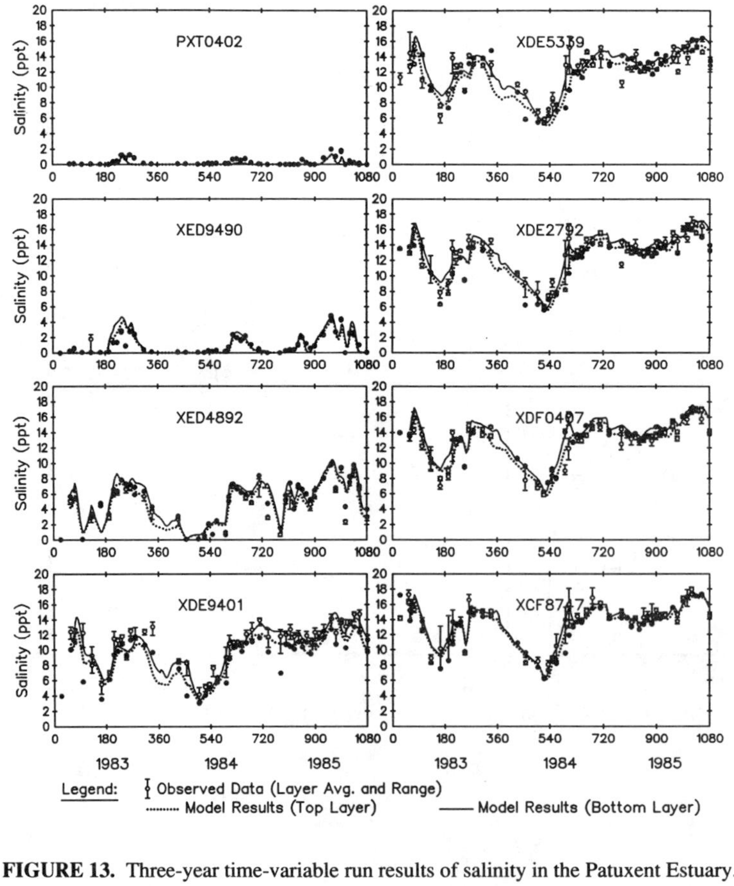

FIGURE 13. Three-year time-variable run results of salinity in the Patuxent Estuary.

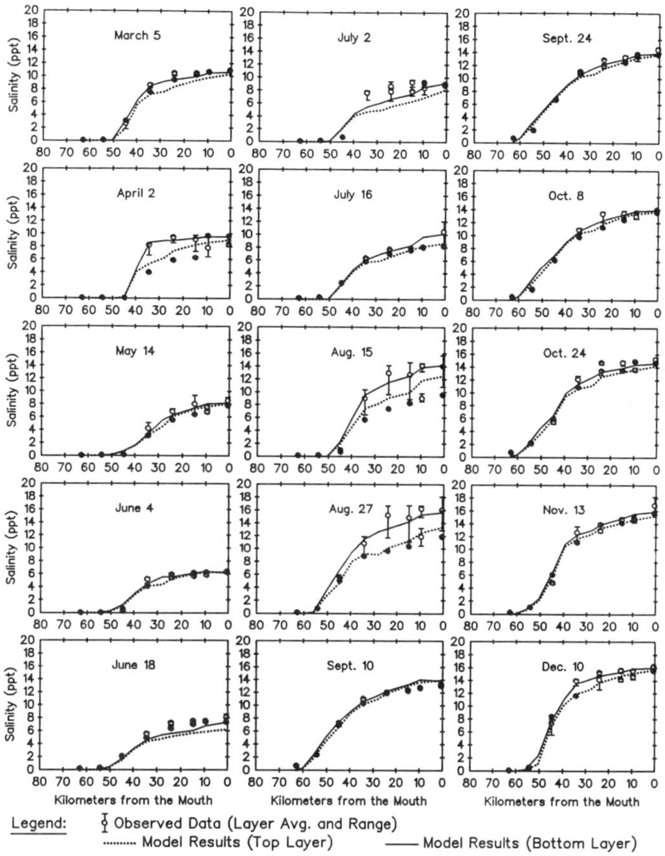

FIGURE 14. Snapshots of spatial salinity profiles in the Patuxent Estuary, 1984.

profiles of salinity are presented in Figure 14 showing the comparisons of model calculations and measured salinity concentrations in the longitudinal and vertical directions in 1984. The results indicate that the model calculations match the data very well in these spatial plots. In general, the two-layer salinity profiles are further apart under high freshwater flows. On the other hand, lower freshwater flows would tend to provide better mixing in the vertical direction with salinity intrusion further inland (see Figure 14).

E. NUMERICAL DISPERSION IN BOX MODELS

The basic task of mass transport calculations is solving the advection-diffusion equation. One of the key calculations in mass transport modeling is quantifying the mass flux across a given cross section. In many box models, the advective mass flux is evaluated using the following approximation derived from the Taylor's series expansion by retaining only the first term for the concentration gradient needed for the advection calculation. In one-dimensional configuration, the approximation for dS/dx (in Equation 3.14), using a backward (upwind) differencing, is therefore:

$$\left(\frac{\partial S}{\partial x}\right)_i = \frac{S_i - S_{i-1}}{\delta x} + \left[\frac{\delta x}{2}\left(\frac{\partial^2 S}{\partial x^2}\right)_i - \frac{(\delta x)^2}{6}\left(\frac{\partial^3 S}{\partial x^3}\right)_i + \ldots\right] \tag{3.46}$$

Higher order terms in the bracket are neglected, resulting in a truncation error. This error shows up as an added dispersive term in the solution of the advection-diffusion equation[33] and can overwhelm the physical dispersion,[53] resulting in physically meaningless computations.

In box models as represented by Equation 3.20, the first term on the right-hand side — $Q_{kj}(\alpha_{kj} C_{mk} + \beta_{kj} C_{mi})$ contributes to the added dispersion that is incorporated into the dispersion coefficient E'_{kj}. The first term in Equation 3.46 is recognized as the backward (or upwind) difference approximation to the spatial concentration derivative. Box models using an upwind finite difference scheme are inherent with numerical dispersion.[33] In box models, the upwind difference scheme is necessary to insure positive solutions in the calculation.

Numerical dispersion for one-dimensional configuration is defined as follows:

$$D_p = u\delta x(1 - C)/2 \tag{3.47}$$

where $C = u\mathrm{d}t/\mathrm{d}x$ is the Courant number. For steady-state box models, Equation 3.47 may be simplified to the following form:

$$D_p = u\delta x/2 \tag{3.48}$$

as the Courant number approaches zero. Thus, the value of numerical dispersion varies with the advective velocity and the size of the segment. Although numerical dispersion could be reduced by carefully segmenting the estuarine system, its persistence could provide difficulty in some circumstances. The available options to control numerical dispersion are reducing the size of the segments and retaining higher order terms in the approximation of the concentration derivative. Increasing the the number of segments may significantly magnify the computational effort, making the numerical task impractical.

There are a number of ways in improving the approximation of the concentration derivative term in the mass transport equation. Leonard[54] described an explicit scheme called QUICKEST (Quadratic Upstream Interpolation for Convective Kinematics with Estimated Streaming Terms). The approximation is based on a conservation form (control volume) formulation of the solute transport equation, so that mass conservation is ensured. Upstream-weighted quadratic interpolation formulae are used for the mass-fluxes through the control volume walls; the formulae are given a strong Lagrangian flavor through the estimated streaming terms, whereby fluxes at the new time level are streamed forward from an accurate estimate of upstream position of the old time level.[54] The inflow and outflow concentrations, or cell wall concentrations, are calculated from a quadratic interpolation function using concentrations at two adjacent elements together with that at the next upstream element. Recently, Abbott and Basco[53] have provided an in-depth evaluation of QUICKEST for solving advection and diffusion equations.

While QUICKEST was originally developed for finite differencing method with constant grid size, it was later adopted for box models with variable box sizes.[55] A case study to demonstrate the benefit of QUICKEST is presented: a two-dimensional, time-variable mass transport calculation for the Patuxent Estuary. Tidally averaged mass transport coefficients (advective and dispersive flows) are used in both analyses. In the two-dimensional calculation, QUICKEST was utilized in two separate directions (longitudinal and vertical) as if each were in a one-dimensional calculation. Temporal salinity plots for 1984 obtained from the calculations using QUICKEST are shown in Figure 15, which should be compared with those without QUICKEST (summarized in Figure 16). The calculated salinity distributions with QUICKEST appear to have increased longitudinal and vertical (two-layer) salinity gradients than those without QUICKEST, an indication that numerical dispersion is reduced by QUICKEST.

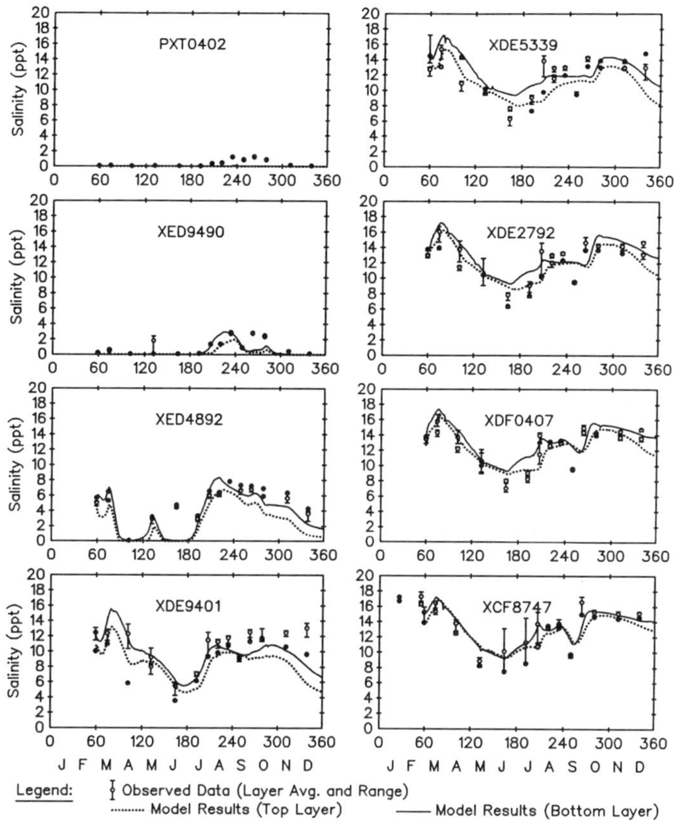

FIGURE 15. Salinity results of mass transport analysis of the Patuxent Estuary, 1984 using QUICKEST scheme.

In box model calculations, upwind differencing is employed to avoid negative solutions and thereby could result in numerical dispersion. Higher order accuracy methods can be used to reduce the numerical dispersion by subtracting out the truncation error. The QUICKEST algorithm offers a good compromise to maintain positive solutions while keeping the numerical dispersion low. Such a reduction of numerical dispersion is significant in a real-time mode with tidal currents at least one order of magnitude higher than the tidally averaged velocity (see Chapter 7).

F. COMPUTATIONAL EFFORT

Almost all the mass transport calculations reported in this chapter were performed on microcomputer systems. Today's rapid advancement on the hardware (execution speed) and software (compiler) technology in microcomputers have made this computing option very attractive. For example, the three-year time-variable salinity simulation for the Patuxent Estuary in a 38-segment configuration takes less than two minutes on a 486 computer at 50 MHz speed. The computational effort involves over 30,000 integration steps in such a three-year run. Additional information on computational effort for water quality modeling is presented at the end of Chapter 4.

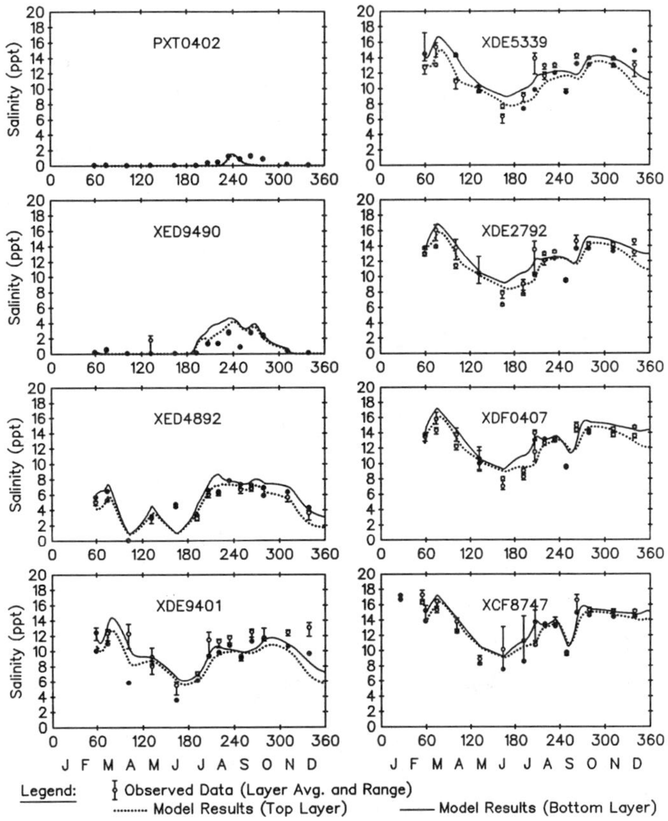

FIGURE 16. Salinity results of mass transport analysis of the Patuxent Estuary, 1984 not using QUICKEST scheme.

REFERENCES

1. **Orlob, G. T.,** Mathematical modeling of estuarial systems, in *Modeling of Water Resources Systems*, Biswas, A.K., Ed., Harvest House, 1972, 87.
2. **Kjerfve, B.,** *Hydrodynamics of Estuaries. Vol. I. Estuarine Physics*, CRC Press, Boca Raton, FL, 1988.
3. **Kjerfve, B.,** *Hydrodynamics of Estuaries. Vol. II. Estuarine Case Studies*, CRC Press, Boca Raton, FL, 1988.
4. **Ippen, A.,** *Estuary and Coastal Hydrodynamics*, McGraw-Hill, New York, 1966, 5.
5. **Thatcher, M. and Harleman, D. R. F.,** A Mathematical Model for the Prediction of Unsteady Salinity Intrusion in Estuaries, R.M. Parsons Laboratory Rep. No. 144, Massachusetts Institute of Technology, Cambridge, MA, 1972.
6. **Shubinski, R. P., McCarty, J. C., and Lindorf, M. R.,** Computer simulation of estuarial networks, *J. Hydraul. Div., ASCE*, 91, 1965.
7. **Blumberg, A. F.,** Numerical model of estuarine circulation, *J. Hydraulics Division, ASCE,* 103, 295, 1977.
8. **Pritchard, D. W.,** Salinity distribution and circulation in the Chesapeake Bay estuarine system, *J. Mar. Res.*, 11, 106, 1952.
9. **Rattray, M., Jr. and Hansen, D. V.,** A similarity solution for circulation in an estuary, *J. Mar. Res.*, 20, 121, 1962.
10. **Boericke, R. R. and Hogan, ?,** An x-z hydraulic/thermal model, *J. Hydraul. Div., ASCE*, 103, 19, 1977.
11. **Bowden, K. F. and Hamilton, P.,** Some experiments with a numerical model of circulation and mixing in a tidal estuary, *Estuar. Coastal Mar. Sci.*, 3, 281, 1975.
12. **Festa, J. F. and Hansen, D. V.,** A two-dimensional numerical model of estuarine circulation: the effects of altering depth and river discharge, *Estuar. Coastal Mar. Sci.*, 4, 309, 1976.

13. **Hodgins, D. O., Osborn, T. R., and Quick, M. G.,** Numerical model of stratified estuary flows, *J. Waterways Port Coastal Ocean Div., ASCE*, 103, 25, 1977.
14. **Hamilton, P.,** On the numerical formulation of a time dependent multi-level model of an estuary, with particular reference to boundary conditions, in *Estuarine Processes*, Vol. II, Wiley, M., Ed., Academic Press, New York, 1977, 347.
15. **Ward, G. H.,** Formation and closure of a model of tidal mean circulation in a stratified estuary, in *Estuarine Processes*, Vol. II, Wiley, M., Ed., Academic Press, New York, 1977, 365.
16. **Olson, P. and Kincaid, C.,** Numerical Model of Patuxent River Estuary Hydrodynamics, draft report submitted by the Johns Hopkins University to Maryland Office of Environmental Programs, 1988.
17. **Wang, D. P. and Kravitz, D. W.,** A semi-implicit two-dimensional model of estuarine circulation, *J. Phys. Oceanogr.*, 10, 441, 1980.
18. **O'Connor, D. J. and Lung, W. S.,** Suspended solids analysis of estuarine systems, *J. Environ. Eng., ASCE*, 107, 101, 1981.
19. **HydroQual, Inc.,** Development of a coupled hydrodynamic/water quality model of the eutrophication and anoxia processes of the Chesapeake Bay, report submitted to U.S. Environmental Protection Agency, Annapolis, MD, 1987.
20. **Johnson, B. H., Kim, K. W., Sheng, Y. P., and Heath, R. P.,** Development of three-dimensional hydrodynamic model of Chesapeake Bay, in *Proc. Estuarine and Coastal Modeling*, Spaulding, M.L., Ed., ASCE, New York, 162, 1990.
21. **Sheng, Y. P., Cook, V., Peene, S., Eliason, D., Scofield, S., Ahn, K. M., and Wang, P. F.,** A field and modeling study of fine sediment transport in shallow waters, in *Proc. Estuarine and Coastal Modeling*, Spaulding, M.L., Ed., ASCE, New York, 113, 1990.
22. **Wang, K. H. and Sheng, Y. P.,** Field validation of a simplified second-order closure model for parameterization of vertical turbulent mixing in three-dimensional hydrodynamic models, paper presented at the Estuarine and Coastal Modeling Conference, Newport, RI, November 16, 1989.
23. **Cheng, R. T. and Smith, P. E.,** A survey of three-dimensional numerical estuarine models, in *Proc. Estuarine and Coastal Modeling*, Spaulding, M.L., Ed., ASCE, New York, 1990, 1.
24. **Bowie, G. L., Mills, W. B., Porcella, D. B., Campbell, C. L., Pagenkopf, J. R., Rupp, G. L., Johnson, K. M., Chan, P. W. H., and Gherini, S. A.,** *Rates, Constants, and Kinetics Formulations in Surface Water Quality Modeling*, 2nd ed., U.S. Environmental Protection Agency, Athens, GA, 1985, 2.
25. **Fischer, H. B., List, E. J., Koh, R. C. Y., Imberger, J., and Brooks, N. H.,** *Mixing in Inland and Coastal Waters*, Academic Press, New York, 1979, 7.
26. **McCutcheon, S. C., Zhu, D., and Bird, S.,** Model Calibration, Validation, and Use, in *Technical Guidance Manual for Performing Waste Load Allocations, Book III Estuaries*, U.S. Environmental Protection Agency, Athens, GA, 1990, 5.
27. **Bedford, K. W.,** *Selection of Turbulence and Mixing Parameterizations for Estuary Water Quality Models*, U.S. Army Engineers Waterways Experiment Station, Miscellaneous Paper EL-85-2, Vicksburg, MS.
28. **Dyer, K. R.,** *Estuaries: A Physical Introduction*, John Wiley & Sons, New York, 1973, 7.
29. **Lung, W. S., Martin, J. L., and McCutcheon, S. C.,** Eutrophication and mixing analysis of embayments in Prince William Sound, Alaska, manuscript accepted for publication in *J. Environ. Eng.*, 1992.
30. **O'Connor, D. J.,** Oxygen balance of an estuary, *J. Sanit. Eng., ASCE*, 86, 35, 1960.
31. **O'Connor, D. J.,** Organic pollution of New York Harbor — theoretical considerations, *J. Water Pollut. Fed.*, 34, 905, 1962.
32. **O'Connor, D. J.,** Estuarine distribution of nonconservation substances, *J. Sanit. Eng., ASCE*, 91, 23, 1965.
33. **Thomann, R. V. and Mueller, J. A.,** *Principles of Surface Water Quality Modeling and Control*, Harper & Row, New York, 1987, 3.
34. **Harleman, D. R. F.,** One-dimensional models, in *Estuarine Modeling: An Assessment*, TRACOR, Inc., prepared for the Office of Water Quality, Environmental Protection Agency, 1971, 3.
35. **Pritchard, D. W.,** Dispersion and flushing of pollutants in estuaries, *J. Hydraul. Div., ASCE*, 95, 115, 1969.
36. **Thomann, R. V.,** *Systems Analysis and Water Quality Management*, McGraw-Hill Book Co., New York, reprinted by J. Williams Book Co., Oklahoma City, OK, 1972.
37. **Officer, C. B.,** Box models revisited, in *Estuarine and Wetland Processes*, Hamilton, P. and Macdonald, K.B., Eds., Plenum Press, New York, 1980, 65.
38. **Chapra, S. and Nossa, G. A.,** Documentation for HAR03 — A Computer Program for the Modeling of Water Quality Parameters in Steady-State Multi-Dimensional Natural Aquatic Systems, U.S. Environmental Protection Agency, Region II, New York, 1974.
39. **Lung, W. S. and O'Connor, D. J.,** Assessment of the Effect of Proposed Submerged Sill on the Water Quality of Western Delta-Suisan Bay, Report submitted by Hydroscience, Inc. to U.S. Army Corps of Engineers, Sacramento District, 1978.
40. **Franco, A. C., III. and Lung, W. S.,** Comparing two estuarine mass transport models, in *Hydraulic Engineering*, Ragan, R.M., Ed., ASCE, New York, 1987, 612.

41. **Officer, C. B.,** *Physical Oceanography of Estuaries (and Associated Coastal Waters),* John Wiley & Sons, New York, 1976, 4.
42. **Officer, C. B.,** Longitudinal circulation and mixing relations in estuaries, in *Estuaries, Geophysics, and the Environment,* National Academy of Sciences, Washington, D.C., 1977, 1.
43. **Lung, W. S. and O'Connor, D. J.,** Two-dimensional mass transport in estuaries, *J. Hydraul. Eng., ASCE,* 110, 1340, 1984.
44. **Lung, W. S.,** Advective acceleration and mass transport in estuaries, *J. Hydraul. Eng., ASCE,* 112, 874, 1986.
45. **Fischer, H. B.,** A feasibility study of multi-layer modeling in Suisan Bay, report prepared for the State of California Department of Water Resources, 1983.
46. **HydroQual, Inc.,** Water quality analysis of the Patuxent River, report prepared for Maryland Office of Environmental Programs, 1981.
47. **Hydroscience, Inc.,** NYC 208 task report on seasonal steady state modeling, report submitted to Hazen & Sawyer, Engineers, New York, 1978.
48. **El-Sabh, M. I. and Murth, T. S.,** Discussion on "two-dimensional mass transport in estuaries," *J. Hydraul. Eng., ASCE,* 113, 799, 1987.
49. **Meade, R. H.,** Transport and deposition of sediment in estuaries, *Memoir 133,* The Geological Society of America, Inc., Boulder, CO, 1972, 91.
50. **Dyer, K. R.,** *Coastal and Estuarine Sediment Dynamics,* John Wiley & Sons, New York, 1986.
51. **Onishi, Y.,** Sediment-contaminant transport model, *J. Hydraul. Div., ASCE,* 107, 1099, 1981.
52. **Lung, W. S.,** Development of a water quality model for Patuxent Estuary, in *Estuarine Water Quality Management: Monitoring, Modelling and Research,* Michaelis, W., Ed., Springer-Verlag, Berlin, 1990, 49.
53. **Abbott, M. B. and Basco, D. R.,** *Computational Fluid Dynamics an Introduction for Engineers,* Longman Scientific & Technical, England, 1989, 5.
54. **Leonard, B. P.,** A stable and accurate convection modelling procedure based on quadratic upstream interpolation, *Comput. Methods Appl. Mechan. Eng.,* 19, 59, 1979.
55. **Dortch, M.S. and Chapman, R. S.,** Integrating time-varyin g, three-dimensional hydrodynamic model output for Chesapeake Bay water quality models, in *Proc. Estuarine and Coastal Modeling,* Spaulding, M.L., Ed., ASCE, New York, 1990, 182.

APPENDIX
SOLVING EQUATION 3.14 WITH EXPONENTIALLY VARYING CROSS-SECTIONAL AREA

Start with the following governing equation:

$$\frac{1}{A}\frac{d}{dx}\left(EA\frac{dC}{dx}\right)-\frac{Q}{A}\frac{dC}{dx}-KC=0 \tag{3.14}$$

The cross-sectional area can be expressed as:

$$A = A_o e^{ax} \tag{A.1}$$

where A_o is the area at $x = x_o$. A is approaching zero as $x = -\infty$. Thus, the cross-sectional area decreases exponentially in the upstream direction. Further,

$$\frac{dA}{dx} = aA_o e^{ax} = aA$$

In addition,

$$\frac{Q}{A} = \frac{Q}{A_o e^{ax}} = U_o e^{-ax}$$

where $U_o = Q/A_o$ and is the tidally averaged velocity at $x = x_o$. U_o decreases exponentially in the downstream direction. Equation 3.14 (with $K = 0$ for a conservative tracer such as salinity) now becomes

$$E\frac{d^2C}{dx^2} + Ea\frac{dC}{dx} - U_o e^{-ax}\frac{dC}{dx} = 0 \tag{A.2}$$

Let

$$w = \frac{dC}{dx}$$

Then

$$\frac{dw}{dx} = \frac{d^2C}{dx^2}$$

Thus,

$$\begin{aligned} &\frac{dw}{dx} + aw - \frac{U_o e^{-ax}}{E} w = 0 \\ &\frac{dw}{dx} - \left(\frac{U_o e^{-ax}}{E} - a\right) w = 0 \end{aligned} \tag{A.3}$$

Equation A.3 is a first-order linear homogeneous differential equation with variable coefficients, which solution is

$$w = C_1 e^{\int\left(\frac{U_o e^{-ax}}{E}+a\right)dx} = C_1 e^{\left(-\frac{U_o}{aE}e^{-ax}-ax\right)} \tag{A.4}$$

Thus,

$$\frac{dC}{dx} = C_1 e^{-\left(\frac{U_o}{aE} e^{-ax} + ax\right)} \tag{A.5}$$

Integrating Equation A.5 yields

$$C = C_1 \int \left(e^{\frac{-U_o}{aE} e^{-ax}} \right) e^{-ax} dx + C_2 \tag{A.6}$$

Let $v = e^{-ax}$. Thus, $dv = -ae^{-ax}dx$. Equation A.6 becomes

$$C = C_1 \int e^{\frac{-U_o}{aE} v} \left(-\frac{1}{a} \right) dv + C_2 = C_1 \left[\frac{aE}{-U_o} \right] e^{-\frac{U_o}{aE} v} \left(-\frac{1}{a} \right) + C_2$$

Thus,

$$C = \frac{C_1 E}{U_o} e^{\frac{-U_o}{aE} e^{-ax}} + C_2 \tag{A.7}$$

The integration constants C_1 and C_2 in Equation A.7 are obtained using the following boundary conditions: at $x = -\infty$, $C = 0$ and at $x = x_o$, $C = C_o$. Applying the first boundary condition yields $C_2 = 0$. Thus, Equation A.7 becomes

$$C = \frac{C_1 E}{U_o} e^{\frac{-U_o}{aE} e^{-ax}} \tag{A.8}$$

and applying the second boundary condition yields

$$C_o = \frac{C_1 E}{U_o} e^{\frac{-U_o}{aE} e^{-ax_o}} \tag{A.9}$$

Solving Equation A.9 for C_1

$$C_1 = \frac{C_o U_o}{E} e^{\left(\frac{U_o}{aE} e^{-ax_o}\right)} \tag{A.10}$$

Therefore,

$$C_1 = \frac{C_o U_o}{E} e^{\frac{U_o}{aE} e^{-ax_o}} \left(\frac{E}{U_o} \right) e^{\frac{-U_o}{aE} e^{-ax}} = C_o e^{\frac{U_o}{aE}\left(e^{-ax_o} - e^{-ax}\right)} \tag{A.11}$$

Chapter 4

WATER COLUMN KINETICS I: DISSOLVED OXYGEN AND EUTROPHICATION

I. INTRODUCTION

Water quality models have been developed for problems in three broad contexts: biochemical oxygen demand (BOD)/dissolved oxygen (DO), eutrophication and nutrients, and toxic substances. This section deals with the kinetics of the first two problem areas which cover a wide spectrum of time and space scales in estuarine systems. Comprehensive descriptions of mathematical formulations and coefficients of water column kinetics for these water quality problems can be readily found in Thomann and Mueller[1] and Bowie et al.[2] Therefore, a detailed treatment of kinetics formulations will not be repeated and only a brief discussion of the kinetics is presented in this chapter. The emphasis of this chapter is the derivation of the kinetic coefficients and technical issues associated with the derivation.

In this chapter, the BOD kinetics are presented first, followed by discussions of technical issues using the latest information available. Two case studies are then presented to further illustrate some of these issues. In the eutrophication topic, a brief description of the kinetics is again presented. Then a number of case studies ranging from seasonal steady-state phytoplankton modeling to time-variable phytoplankton modeling are described. A number of subtle issues related to water quality management are also explored. Finally, tools for eutrophication modeling analysis are discussed.

II. BOD/DO KINETIC PROCESSES IN THE WATER COLUMN

Figure 1 shows the major kinetic processes in the water column:

- Reaeration
- Carbonaceous deoxygenation
- Nitrogenous deoxygenation (nitrification)
- Algal Photosynthesis and Respiration
- Sediment oxygen demand
- Settling and deposition of organic material

A. BIOCHEMICAL OXYGEN DEMAND

Biochemical oxygen demand (BOD) is a measure of the oxidizable matter by biochemical processes. That is, it combines the effects of several oxygen consuming processes into one variable. The BOD of a wastewater is the amount of oxygen required by the stabilization of organic matter. The test is conducted at an incubation temperature of 20°C and is usually run for five days, at the end of which time the amount of oxygen used is determined. The test sample is seeded with raw wastewater, treated effluent, or river water. Samples taken directly from natural waters usually contain sufficient bacteria to carry on the oxidation and seeding is not required. The oxidation process is generally carried out in two stages: carbonaceous and nitrogenous. The first stage is accomplished by the saprophytic organisms, those that derive their energy from the breakdown of organic compounds, and the second by the autotrophic bacteria that require the simple inorganic compounds. Each stage is characterized by two steps: synthesis and respiration. In the carbonaceous stage, the energy required for synthesis is obtained from the destruction of complex organic compounds liberating carbon dioxide and water. After the

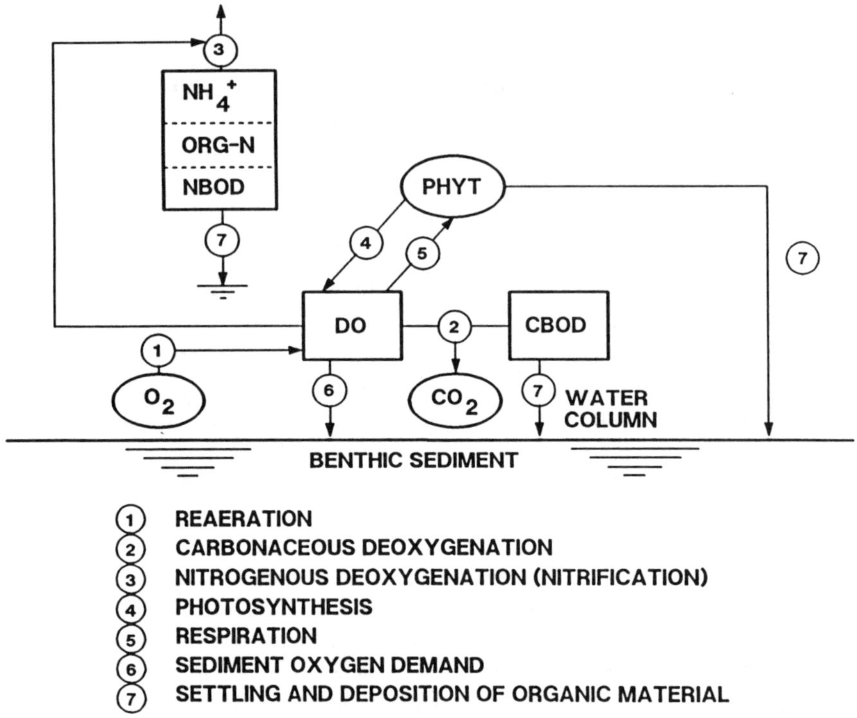

FIGURE 1. BOD/DO kinetic processes in water column.

organic matter has been converted to bacterial cells, the endogenous respiration of the synthesized organisms occurs, also yielding carbon dioxide and water and usually ammonia. If insufficient nutrients (nitrogen and phosphorus) are present they must be added initially in order that the process may proceed at a reasonable rate. In the BOD test, there is a pronounced lag between the carbonaceous oxidation and the nitrification step, the latter following by as much as ten days. The lag is less for the treated (stabilized) wastewaters and is in the order of one or two days for highly treated effluents. In an estuary, the two stages frequently proceed simultaneously although there may be lags in the nitrification stage in highly polluted streams, or those with low dissolved oxygen.

By U.S. convention, many BOD measurements are only conducted for five days. In fact, wastewater discharge permits (NPDES) are written in terms of 5-day BOD by regulatory agencies. In addition, many of the tests are run with an inhibitor of nitrification so that the test measures the oxidation of carbonaceous material only. When total BOD is measured after five days (an inhibitor is not used), these tests are designated as BOD_5. When the five-day test employs a nitrification inhibitor, the results are designated as $CBOD_5$. More and more frequently, long-term tests of 20 to 30 days are employed to measure ultimate BOD designated as $CBOD_u$ to reflect the potential strength of the oxygen consumption. Some paper and pulp mill wastewater samples are analyzed for much longer periods (in excess of 100 days), but these measurements are not very useful in estuaries where the time of travel from the waste source to the dissolved oxygen sag is only a a few days or weeks where the estuary is diluted by tributaries within a few days.

The nitrogenous stages of the BOD test include conversion of organic nitrogen to ammonia and the subsequent oxidation of ammonia. Many wastewaters contain both organic nitrogen, such as urea, and/or ammonia. The former is hydrolyzed to ammonia, under aerobic or anaerobic conditions, without the utilization of oxygen. Ammonia is subsequently oxidized through nitrite to nitrate by the organisms nitrosomonas and nitrobactor, respectively.

The most common approach to modeling carbonaceous deoxygenation and nitrification is to use first-order kinetics. That is, the rate of accumulation or depletion is linearly dependent on the amount of carbon or nitrogen available in a specific pool. Because of the ease of measuring organic nitrogen, ammonia, nitrite, and nitrate, modeling nitrification involves a mass balance and description of the decay of CBOD and each nitrogen species. Nitrification is best simulated separately from CBOD as a cascade process involving hydrolysis of organic nitrogen, oxidation of ammonia, and oxidation of nitrite. In some models, the intermediate step of nitrite oxidation is lumped into the overall oxidation of ammonia to nitrate but only a little computational efficiency is gained. Some earlier models lumped the cascade process into a single process and combining all nitrogenous oxygen demands as NBOD. Modeling NBOD and CBOD as separate demands is *not* as useful as modeling CBOD, ammonia, nitrite, and nitrate, which is widely used in wasteload allocation studies.

B. REAERATION

In general, oxygen may be removed from or added to water, by various physical, chemical, or biological reactions. If oxygen is removed from the water column and the concentration drops below the saturation level, the deficiency is made up by the transfer of the gas from the atmosphere through the surface into the estuary. If oxygen is added and the water column concentration is greater than the saturation level, the supersaturation is reduced by the transfer from the estuary to the air. Such interactions between the gas phase and liquid phase are driven by the partial pressure gradient in the gas phase and the concentration gradient in the liquid phase. The rate of transfer to be quantified in the stream BOD/DO modeling analysis is expressed as:

$$\frac{dC}{dt} = K_a(C_s - C) \tag{4.1}$$

where

C_s = saturation concentration of dissolved oxygen
C = dissolved oxygen concentration in estuary
K_a = reaeration coefficient (d^{-1})

If the estuary is undersaturated, $C < C_s$, oxygen passes into water and if supersaturated, $C > C_s$, oxygen escapes from the water.

Two parameters, K_a and C_s, are needed for Equation 4.1. In general, oxygen transfer in natural waters depends on:

- Internal mixing and turbulence due to velocity gradients and fluctuation
- Temperature
- Wind mixing
- Waterfalls, dams, rapids
- Surface films

Because of the complexity, theoretical calculations of reaeration coefficient is difficult. On the other hand, no simple technique has been developed for accurately determining reaeration coefficients. Measuring methods, which have been reported in the literature, fall into two categories: using mass balance to back-calculate the reaeration coefficient and tracer methods. In field investigations, the general procedure was to measure the reaeration coefficient indirectly by changes in dissolved oxygen under closely controlled field conditions. Additional discussion on determining reaeration rates in estuaries is presented in Section III.E.

C. SEDIMENT OXYGEN DEMAND

Benthic decomposition of organic material is defined as the stabilization of the volatile suspended solids, which have settled to the stream bed. These deposits are stabilized by the biological activity of many different organisms. As these organic materials are associated with suspended solids, the discharge of settleable waste components may form a sludge blanket below a wastewater outfall. After a period of time, organic materials may accumulate.

The demand of oxygen by sediment and benthic organisms can, in some instances, be a significant fraction of the total oxygen demand especially in deep waters. The effects may be particularly acute during low flow and high temperature conditions. Decomposition of organic matter and respiration of resident invertebrates form the major oxygen demands from the sediment. Although these processes are distinct, they are typically quantified together since *in situ* measurements lump together oxygen uptake, and separation would result in additional model complexity.

Due to its complexity, it is difficult to estimate the sediment oxygen demand analytically and independently. *In situ* measurements of sediment oxygen demand are usually conducted using a chamber of the bottom of the stream. Continuous measuring of oxygen uptake over a certain period of time provides data to derive the oxygen consumption rate. In some cases, samples of river sediments (undisturbed) are taken to the laboratory for measuring the oxygen uptake of the bottom muds. The amount of oxygen used over the test period is calculated as g-O_2/m_2-d. In modeling analysis, sediment oxygen demand is typically formulated as zero-order process as follows:

$$\frac{dC}{dt} = SOD / H \tag{4.2}$$

where

dC/dt = rate change of oxygen concentration (g-O_2/m^3-d)
SOD = sediment oxygen demand (g-O_2/m^2-d)
H = average river depth (m)

Like many other BOD reaction coefficients, the SOD values could be determined by model calibration if direct measurements from the field are not available. The difficulty arises when SOD values need to be predicted for future conditions. In recent years, credible interactive sediment models are appearing to independently quantify the oxygen uptake rates of sediments. For example, Di Toro[3] has developed a SOD model based on diagenesis of particulate organic materials to predict the production of H_2S, NH_4^+, PO_4^{-3}, and Si. Such a framework explains some but not all of the processes associated with SOD and is still being tested.

D. ALGAL PHOTOSYNTHESIS AND RESPIRATION

Through photosynthesis and respiration, phytoplankton, periphyton, and rooted aquatic plants (macrophytes) can significantly affect the dissolved oxygen levels in the water column. As phytoplankton growth requires sunlight and nutrients, quantifying photosynthetic oxygen production would need to address phytoplankton — nutrient dynamics which in turn is related to the availability of nutrient and recycling of nutrients. That is, phytoplankton and nutrient should be modeled concurrently to address this problem. However, in many simple estuarine BOD/DO models, the oxygen production rate due to photosynthesis and consumption rate due to respiration are assigned, thereby decoupling the calculation from the phytoplankton — nutrient dynamics. In this section, such an approach is presented while the full discussion of phytoplankton — nutrient dynamics is found in the eutrophication analysis.

In an estuarine water quality model, the daily average oxygen production due to photosynthesis and reduction due to respiration may be formulated as follows:

$$\frac{dC}{dt} = P - R \tag{4.3}$$

in which

dC/dt = rate of change of oxygen concentration (mg-O_2/l-d)
P = average gross photosynthesis production (mg-O_2/l)
R = average respiration (mg-O_2/l)

Note that R is considered to be plant respiration only, excluding microbial respiration for the CBOD and NBOD. Three methods are used in estimating P and R in estuarine water quality modeling studies:

- Light and dark bottle measurements of dissolved oxygen
- Estimation from observed chlorophyll *a* levels
- Measurements of diurnal variations of dissolved oxygen concentrations

The light and dark bottle method measures the change (increase) in dissolved oxygen as an estimate of photosynthesis gross dissolved oxygen production minus the dissolved oxygen utilized in respiration, the net production of dissolved oxygen. The dark bottle provides an estimate of the total dissolved oxygen consumption, since the plants are not photosynthesizing. Any oxygen consumption by bacteria (BOD) must be quantified independently and subtracted from total respiration.

The second method is addressed to the following problem: given the concentration of phytoplankton in an estuary, estimate the average daily oxygen production. A technique has been developed by Di Toro[5] and its brief derivation can also be found in Thomann and Mueller.[1] In summary, the following equations are used:

$$P = 0.25\ Chla \tag{4.4}$$

and

$$R = 0.025\ Chla \tag{4.5}$$

in which *Chla* is chlorophyll *a* concentration in μg/l.

Di Toro[5] developed an analytical method to calculate P based on the measured diurnal dissolved oxygen range:

$$P = \frac{fK_a\left(1 - e^{-K_aT}\right)}{\left(1 - e^{-K_afT}\right)\left(1 - e^{-K_aT(1-f)}\right)}\Delta \tag{4.6}$$

in which

f = photo period ($0 < f < 1$)
K_a = reaeration coefficient (d^{-1})
T = 1 d
Δ = diurnal dissolved oxygen range (mg/l)

Note that Equation 4.6 can be used to estimate the diurnal range of dissolved oxygen with an estimate of P from the first two methods.

III. IMPORTANT TOPICS ON BOD/DO KINETICS

In a recent paper, Thomann[6] stated, "The dissolved oxygen problems, connected intimately with primary productivity and sediment effects, in spite of the long history, tend to be considerably more complex than generally believed." In addition, the BOD itself has rendered water quality modelers an equal, if not more, amount of difficulties, some of which probably have not received their rightful share of attention. Two factors contribute to this situation. First, BOD is a surrogate; that is, it is simply an indicator, which was derived out of convenience. Second, using first-order equations to formulate the BOD/DO kinetics is straightforward, yet highly empirical. Thus, assigning the kinetic coefficients is difficult, if not totally impossible.

As shown in Figure 1, the important kinetic coefficients are BOD deoxygenation coefficient (K_d), nitrification coefficient (K_n), and estuarine reaeration coefficient (K_a). Their values may not be properly quantified for model projections under future conditions. There are a number of technical issues associated with these kinetic coefficients in estuarine BOD/DO modeling:

- Effluent characteristics
- Nitrification in estuaries
- BOD rates before and after treatment improvement
- Contribution of algae biomass to CBOD
- Algal contribution to dissolved oxygen
- Estuarine reaeration coefficient

A. EFFLUENT CHARACTERISTICS

One of the important factors in dissolved oxygen modeling is $CBOD_u$ to $CBOD_5$ ratio in the effluent. This ratio is used to estimate the ultimate oxygen demand from the measured effluent $CBOD_5$. It is also used to translating model calculated $CBOD_u$ into effluent limits ($CBOD_5$) in NPDES permits. As expected, this factor is closely related to the changes in the effluent quality, which is the result of waste treatment upgrade. In general, the $CBOD_u$ to $CBOD_5$ ratio increases with the treatment level.

Historically, the ratio of $CBOD_u$ to BOD_5 has been assumed as 1.5 for secondary or less treated effluent.[7] It is known that the $CBOD_u$ to BOD_5 ratio is a function of the level of treatment and the associated degradability of the wastewater. Thus, higher ratios are expected and have been observed for higher levels of treatment since the *CBOD* remaining in more highly treated effluents degrades more slowly than that in less treated wastewaters. In a recent study by Leo et al.,[8] data from 144 POTWs were examined to yield an average $CBOD_u/BOD_5$ of 2.47 (with a standard deviation of 1.5) and $CBOD_u/CBOD_5$ of 2.84 (with a standard deviation of 1.2). These values are substantially higher than the ratio of 1.5 which was widely used in the past for primary effluents and some marginal secondary effluents. Some of the $CBOD_u$ to BOD_5 ratios are compiled in Table 1. EPA[9] now suggests that a $CBOD_u$ to $CBOD_5$ ratio of 2.5 to 3.0 be used for plants with nitrification. It must be emphasized that this value is presently based on a limited data set, and that additional treatment plant effluent data is needed to gain greater confidence in the suggested values. In addition, the standard deviation indicates a need for site-specific measurements of $CBOD_u$.

Whenever possible, data from existing plant or pilot plant effluents should be collected to assist in the selection of an appropriate conversion ratio. Caution must be exercised, however, when using data from an existing plant that has a level of treatment significantly lower than that of a proposed plant. In this case, the existing data should not be blindly applied when selecting the appropriate ratio; it should merely be used as a guide. A sensitivity analysis of the ratio and its implications on the final treatment decision to be made can help the modeler determine the relative importance of gathering additional data.

The $CBOD_u$ to $CBOD_5$ ratio depends on the bottle CBOD decay rate (K_1) of the wastewater. As the level of treatment increases, the rate at which the CBOD oxidizes decreases. The following equation illustrates this relationship:

$$\frac{CBOD_u}{CBOD_5} = \frac{1}{1 - e^{-5K_1}}$$

TABLE 1
$CBOD_u/CBOD_5$ — Wastewater Treatment Plants in James River Basin

Plant	Treatment level	Flow (mgd)	$CBOD_u/CBOD_5$
Army Base	Secondary[a]	11.2	5.00
Boat Harbor	Secondary[a]	16.1	2.67
Chesapeake/Elizabeth	Secondary[a]	11.2	2.44
James River	Secondary[a]	6.8	5.20
Lamperts Point	Primary[b]	14.1	1.39
Williamsburg[c]	Secondary[a]	7.98	6.29
Richmond	Secondary[a]	54.9	3.01
Falling Creek	Secondary[a]	7.2	3.00
Proctor's Creek	Secondary[a]	2.8	1.90
Hopewell	Secondary[a]	22.9	4.00

[a] Activated sludge process.
[b] With phosphorus removal.
[c] With large flows from Anheuser Busch (high BOD conc. in effluent, 700 mg/l).

Nitrification in the treatment plant effluent plays an important role in BOD/DO modeling of estuaries. It is known that effluents from secondary facilities often contain populations of nitrifying organisms sufficient to utilize a significant amount of oxygen during the regular five-day incubation period.[10] Thus, nitrification can cause BOD_5 values to exceed effluent limits set for treatment facilities.[11] It was estimated that nearly 60% of the compliance violations nationwide may result from nitrification occurring in the BOD_5 test, rather than from improper design or operation. The U.S. EPA has since changed the rule on secondary treatment to allow use of the $CBOD_5$ test as a measure of compliance.

It is therefore essential that both $CBOD_5$ and BOD_5 be measured of the wastewater for modeling use. There are some speculations that when the nitrification suppressor is contaminated or metabolized, addition of the suppressor to previously inhibited effluent samples may result in increased oxygen depletion. Such an effect could be so significant that the oxygen consumption of a sample inhibited with the suppressor may be greater than the consumption of the sample without the suppressor added.

One possible source of nitrifying bacteria are the effluents of some secondary treatment plants. The work by Hall and Foxen[11] pointed out that the possible presence of nitrifying bacteria in the secondary and highly treated effluents and, therefore, the strong possibility of nitrification in the receiving water. The general understanding is that improving treatment will increase the nitrification rate in the receiving water.[9]

Table 2 summarizes the $CBOD_u$ to $CBOD_5$ ratio for different treatment levels. Also shown in Table 2 are POTW effluent concentrations of BOD_5, $CBOD_5$, and ammonia. Leo et al.[8] indicated these BOD_5 and $CBOD_5$ concentrations are significantly different based on a t test at a 90% confidence level. This information reinforces the findings by Hall and Foxen[11] that significant nitrification is occurring during the BOD test for many secondary treatment POTWs. A similar difference between effluent BOD_5 and $CBOD_5$ concentrations is not observed for secondary facilities that are removing phosphorus from their effluents. This may be because phosphorus removal unit processes also remove nitrifying bacteria from the effluent stream.[8] These data also show ammonia nitrogen effluent concentrations for POTWs designed to nitrify (secondary plants with nitrification) average 1.0 mg/l or less with a standard deviation of about 1.0 mg/l. A summer effluent ammonia concentration for nitrifying POTWs therefore of about 1.0 mg/l appears to be a reasonable estimate for planning purpose.[8]

The relationship between the ratio of $CBOD_u$ to $CBOD_5$ and K_1 the ratcs behaves differently for POTW and industrial effluents. For example, the carbonaceous materials in POTW effluents decay relatively rapidly. Industrial effluents often decay much more slowly, thus exerting a

TABLE 2
Summary of POTW Effluent Characteristics

Treatment type	BOD_5 (mg/l)[a]	$CBOD_5$ (mg/l)[a]	Ammonia-N (mg/l)[a]	$CBOD_u/CBOD_5$
Primary	101.0 (±21.2)[b]			
Trickling Filter	41.2 (±27.8)		16.6 (±12.2)	1.5
Secondary	19.1 (±16.3)	10.3 (±6.4)	8.9 (±6.3)	2.0
Secondary + P-Removal	16.2(±14.0)	14.6 (±9.3)	7.9 (±8.9)	2.3
Secondary + Nitrification	11.5 (±11.8)	4.8 (±3.9)	1.0 (±1.4)	2.8
Secondary + P-Removal + Nitrification	13.6 (±18.6)		0.9 (±0.7)	2.8
Secondary + Nitrification + Filters	3.9 (±2.0)		4.8 (±8.2)	3.2

[a] From Leo et al.[8]
[b] Standard deviation.

longer-lasting but weaker demand on ambient dissolved oxygen. If inappropriately high oxidation rates are applied to these more refractory effluents, the model may incorrectly predict severe dissolved oxygen consumption. It is generally accepted that many industrial wastes are less degradable than conventional sanitary wastewater effluents. As such, high $CBOD_u$ to $CBOD_5$ ratios and low CBOD decay rates are common for industrial wastes.

The $CBOD_u$ to BOD_5 ratio of industrial wastewater is highly dependent on the type of industry, manufacturing processes, treatment schemes or operation, measurement techniques, etc. As such, the ratio depends on the effluent characteristics, such as the $CBOD_u$ and BOD_5 concentrations, which in turn are a function of the type of the mills. Figure 2 displays the relationship between the BOD_5 concentration and the $CBOD_u$ to BOD_5 ratio for various pulp and paper mills. The limited data gathered indicates a high degree of variability for BOD_5 and $CBOD_u/BOD_5$.

B. NITRIFICATION IN ESTUARIES

Nitrification in the receiving water is another difficulty associated with BOD/DO modeling in light of the fact that it usually lags behind the carbonaceous BOD deoxygenation. This presents a problem in model projection for future conditions where carbonaceous BOD loads from the treatment plants are reduced. A good example of this difficulty was shown by O'Connor et al.[12] on the water quality modeling of the Delaware Estuary. As a result, assigning proper nitrification rates for a wasteload allocation study (WLA) is not an easy task because WLA calculations are not simply model simulations using the once verified model. Rather, they usually have difficulties in assigning reasonable rates under a somewhat unknown future condition. Large concentrations of algae, either suspended or attached, markedly affect nitrification rate, K_n. Evaluation of K_n based solely on the loss of ammonia would result in the overestimation of K_n where significant algal effects occur. Algae consume ammonia as a nutrient. Therefore, a K_n rate calculation based only on the loss of ammonia would include uptake of ammonia by algae as well as ammonia oxidation. In some cases, use of the rate of increase of nitrate is a better approach for estimating K_n because nitrate increase results directly from oxidation of ammonia in the stream. However, as a cautionary note, under some conditions, algae can uptake nitrate as well as ammonia. Therefore, the K_n rate derived from nitrate data would represent the minimum K_n.

The effect of algae in determining K_n values can be demonstrated in Figure 3, showing ammonia and nitrate concentration profiles downstream from a treatment plant where a nitrification process is being proposed. Significant declines of ammonia concentrations are shown between river miles

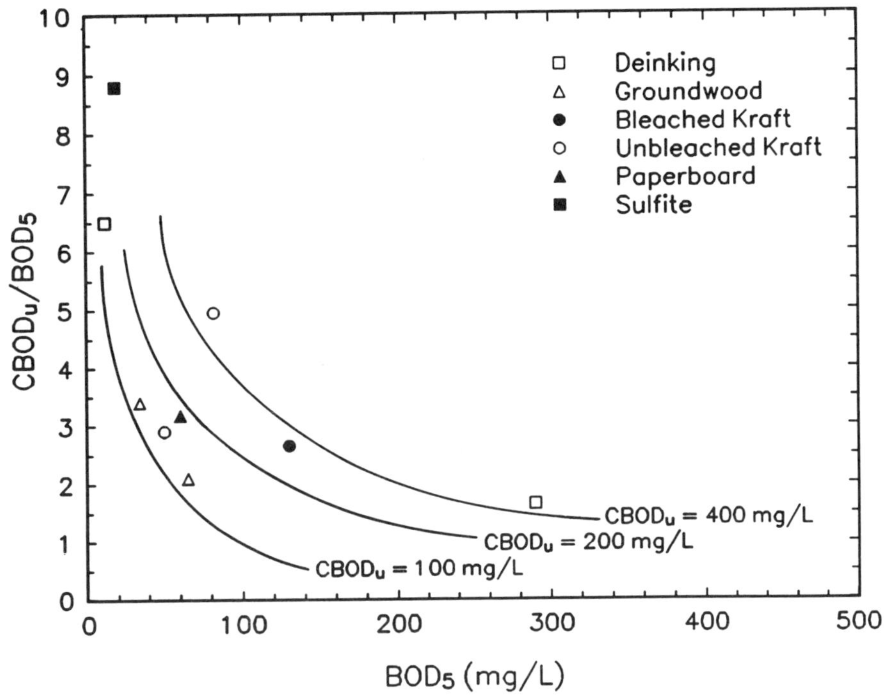

FIGURE 2. *$CBOD_u$* to *BOD_5* ratios for paper and pulp mill effluents.

12 and 8 and from miles 8 through 6. A K_n value of 1.0 d^{-1} can be calculated based on the ammonia profile. Although no chlorophyll *a* data were available, dissolved oxygen data show diurnal fluctuation of 4 to 6 mg/l, which indicates the presence of algae or periphyton capable of uptaking a significant amount of ammonia. Thus, the apparent 1.0 d^{-1} K_n rate likely reflects ammonia uptake by algae and aquatic plants, as well as ammonia oxidation.

Figure 3 also shows relatively constant nitrate concentrations throughout the stream reach. As a result, the 1.0 d^{-1} K_n rate based on the reduction of ammonia concentrations would exceed by five times that derived from the nitrate data. As such, uptake of ammonia by algae instead of ammonia oxidation may have caused the reduction in ammonia concentration.

Another important factor in determining the K_n rate involves the time of the year these rates are determined and applied. Although ammonia and nitrate data may indicate relatively high K_n rates during the summer months, K_n rates for the same estuarine system may be negligible during the winter months, and even during transition periods such as April, May, and June, and September, October, and November. Since construction and operating costs can increase markedly for nitrification at wastewater temperatures below 20°C, K_n rates should be based on site-specific data, particularly if a nitrification process is proposed for such periods.

While improved wastewater treatment is expected to increase the nitrification rate in the receiving water, there is the possibility that nitrification may be inhibited in the receiving water. Several types of inhibition which are commonly considered in the nitrification reaction are

- Toxicity — the presence of something lethal (e.g., total residual chlorine from the chlorination process) to one or both of the nitrifying bacteria
- Chemical limitation — the lack of ammonia, nitrite, oxygen, or alkalinity could prevent the nitrifying bacteria from carrying out their biochemical reactions
- Environmental limitations — water temperature, pH, etc., which are not optimal for bacterial growth

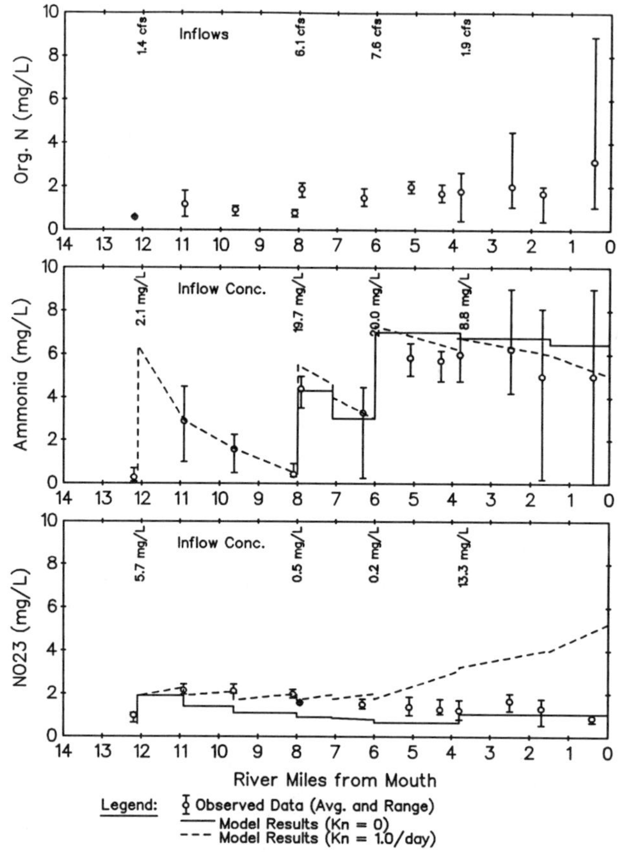

FIGURE 3. Algal effect on nitrification rate.

Results from nitrifying bacteria studies would provide necessary information to address this issue and to quantify the rates for modeling use.

C. BOD RATES BEFORE AND AFTER TREATMENT IMPROVEMENT

In a comprehensive analysis, Leo et al.[8] surveyed a number of case studies on the improvement of the receiving water quality following upgrade or expansion of waste treatment facilities. They found significant differences in BOD kinetic rates in the receiving water before and after the improvement. Three sets of oxidation coefficients show changes after treatment was upgraded while three sets of rates remain the same. Carbonaceous oxidation rates which changed after upgrading, are reduced on the average of 60%. Further, as discussed earlier, the ratio of $CBOD_u$ to $CBOD_5$ in the receiving water tend to increase as the K_d decreases because of treatment improvement.

Although there are guidelines in the literature on exactly how decay rates decrease as treatment levels are upgraded, the K_d cannot be precisely determined prior to the improvement. The best way to verify or confirm this new K_d rate is through a model post-audit analysis. For example, a wasteload allocation study for the upper Mississippi River used a conservative estimate of K_d of 0.25 d^{-1} at 20°C to allocate the CBOD and ammonia loads for the Metropolitan Plant in St. Paul, Minnesota. A model postaudit analysis using the data collected following the treatment upgrade showed that the K_d rate had reduced substantially to 0.073 d^{-1} at 20°C.

While the K_d rate for CBOD usually reduces following the treatment improvement, the instream nitrification rate may increase following the nitrification at the wastewater treatment plant. For example, Leo et al.[8] found that the nitrification rates in Hurricane Creek doubled and

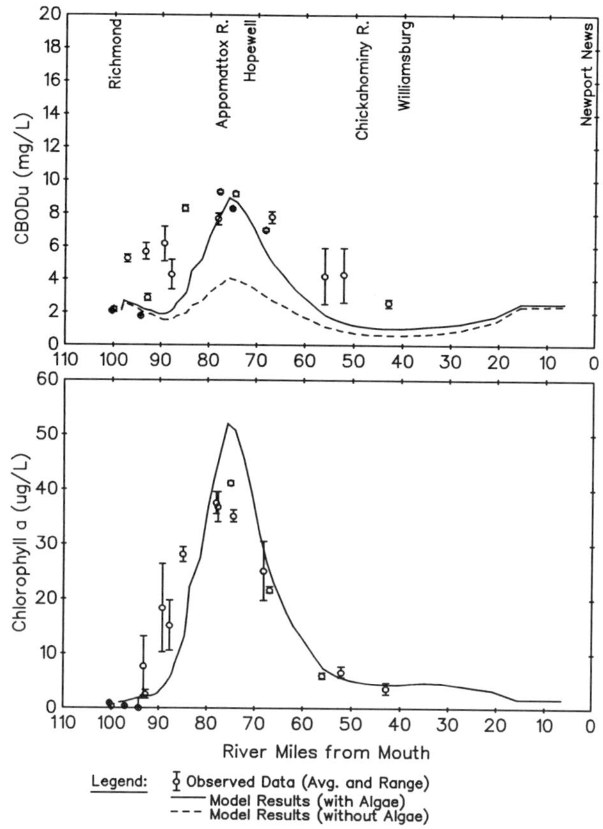

FIGURE 4. Algal contribution to CBOD in James River Estuary — September 1983.

rates in the Clinton River increased from 0.0 d^{-1} to over 2.0 d^{-1}. In general, a bioassay study on the nitrifying bacteria in the receiving water is the best test to discern this issue and to quantify the nitrification rate.

D. CONTRIBUTION OF ALGAE BIOMASS TO CBOD

In estuaries with significant algal biomass, the addition of CBOD recycled from phytoplankton biomass could be considerable. In modeling the upper James Estuary, Lung[13] noted that the CBOD curve without algae is consistently below the observed BOD data. Since the river water samples contained the concentrations of phytoplankton found in the river, the results reflect two components of oxygen demand. The first is the demand created by oxidation of organic waste material and the second is the combined demand created by the respiration of living algae and oxidation of dead algae contained in the sample. The significant increase in measured $CBOD_{40}$ between river miles 70 and 90 is primarily due to the large algae component of oxygen demand in samples taken in this area (Figure 4). The $CBOD_{40}$ measured in the upper James Estuary represent complete decomposition of organic carbon in the unfiltered sample.

A similar result was also obtained from the study of the Delaware Estuary.[14] In the Delaware Estuary, the added oxygen demand from the algal biomass in the water column match the observed BOD_5 data reasonably well. Matching the model calculated CBOD with the measured values depends on the carbon to chlorophyll *a* ratio, which is not usually independently measured and is therefore a parameter with a great degree of uncertainty.

Algae can affect the CBOD data used to calculate the CBOD decay rate. Algal respiration and decay in the CBOD bottle can cause higher measured CBOD values and thus higher K_1 rates compared to samples without algae. In addition, if the concentration of suspended algae (i.e., algae that would not get into the BOD bottle) is not constant in the estuarine system below the discharge, the measured CBOD concentrations would not indicate a defined rate of decay (K_d). Where the concentration of suspended algae increases in the estuary, there may be a net increase of measured CBOD below the discharge.

Since the presence of algae complicates the estimation both K_1 and K_d rate, a laboratory procedure to minimize these effects is necessary. Such a procedure involves measuring filtered CBOD together with total CBOD at several locations downstream of the discharge. The oxygen demand resulting from settleable (i.e., filtered) organics is then accounted for separately in establishing the K_r (settling) and SOD rates. However, even with filtered and unfiltered CBOD data, it is difficult to select model coefficients for calibration and projection modeling for algae dominated streams. Where large concentrations of algae occur in the receiving water, a range of K_d rates should be estimated based on filtered and unfiltered CBOD data. Some general rules of thumb follow:[9]

- Algal impacts on K_d occur wherever high concentrations of chlorophyll *a* or large diurnal dissolved oxygen fluctuations occur.
- 10 μg/l of chlorophyll *a* will increase $CBOD_u$ the concentration by 1 mg/l above that without algae. Rough estimates may be obtained from multiples of this relationship; i.e., 100 μg/l of chlorophyll *a* may increase the $CBOD_u$ concentration by 10 mg/l.
- If the estuarine system is effluent-dominated with most of the CBOD from the discharge rather than algae, do not filter. If the system is not effluent-dominated and most of the CBOD is from an algae, run filtered and unfiltered samples.

E. ESTUARINE REAERATION COEFFICIENT

In estuaries, reaeration has been computed as a function of bottom shear stress or evaluated as an empirical constant. O'Connor[15] used the following equation to calculate the reaeration rate in estuaries:

$$K_a = \frac{(D_L U_o)^{1/2}}{H^{3/2}} \tag{4.7}$$

where

K_a = the reaeration coefficient (d^{-1})
U = mean tidal velocity over a complete cycle (m/d)
D_L = molecular diffusivity of oxygen (m^2/d)
H = average depth (m)

Harleman et al.[16] proposed the following equation:

$$K_a = 10.86 \frac{V^{0.6} H\, W_T}{H^{1.4} A} \tag{4.8}$$

where

V = tidal velocity (ft/sec)
H = average depth (ft)
W_T = top width (ft)
A = cross-sectional area (ft^2)

The constant 10.86 is the recommended value, but can be changed as discussed by Harleman et al.[16]

In the past, wind effects are seldom considered. In a recent review of estimating reaeration rates, Cerco[17] pointed out the importance of wind effect on evaluating the reaeration coefficient for estuaries and coastal waters. O'Connor[18] developed a relation between the transfer mass coefficient of slightly soluble gases (such as oxygen) and wind velocity. For hydrodynamically smooth flow, viscous conditions prevail in the liquid sublayer that controls transfer, and the transfer is affected solely by molecular diffusion. In fully established rough flow, turbulence extends to the surface and turbulent transfer processes dominant. In the transition region between smooth and rough flow, both transfer mechanisms contribute. Based on the physical behavior in the smooth and rough layers K_L is then developed by O'Connor[18] as

$$\frac{1}{K_L} = \frac{1}{\left[\dfrac{D}{v_w}\right]^{2/3} \dfrac{\kappa^{1/3}}{\Gamma(u^*)} \dfrac{\rho_a u^*}{\rho_w}} + \frac{1}{\left[\dfrac{Du^*}{\kappa z_0 u^*} \dfrac{P_a \nu_a}{\rho_w \nu_w}\right]^{1/2}}$$

where

D = molecular diffusivity
v_a = kinematic viscosity of air
v_w = kinematic viscosity of water
k = the Von Karmen constant
ρ_a = density of air
ρ_w = density of water
u^* = shear velocity

$z_0 u^*$ is related to wind speed and drag coefficient; and the parameters characterizing the drag coefficient can be found in O'Connor.[18]

In the eutrophication model for the Potomac Estuary, Thomann and Fitzpatrick[19] used the following equation for reaeration coefficient:

$$K_a = 13\frac{V^{0.5}}{H^{1.5}} + \frac{3.281}{H}(0.728W^{0.5} - 0.317W + 0.0372W^2) \tag{4.9}$$

in which

V = depth averaged velocity (ft/sec)
H = depth (ft)
W = wind speed (m/sec)

It should be pointed out that the first part of Equation 4.9 is the original reaeration equation for streams by O'Connor[15] and the second part of the equation is derived from an equation for calculating reaeration coefficients for lakes and reservoirs.[1]

A compilation of gas exchange rates measured in San Francisco Bay with those for other wind-dominated systems updates previous compilations and yields an equation for predicting gas exchange by Hartman and Hammond:[20]

$$K_L = 34.6\, R_v\, (D_{m20})^{0.5}\, (U_{10})^{1.5} \tag{4.10}$$

where

R_v = ratio of the kinematic viscosity of pure water at 20°C to the kinematic viscosity of water at the measured temperature and salinity
D_{m20} = molecular diffusivity of dissolved oxygen (cm²/sec)
U_{10} = wind speed (m/sec) at 10 m above the surface
K_L = liquid phase transfer coefficient (m/d)

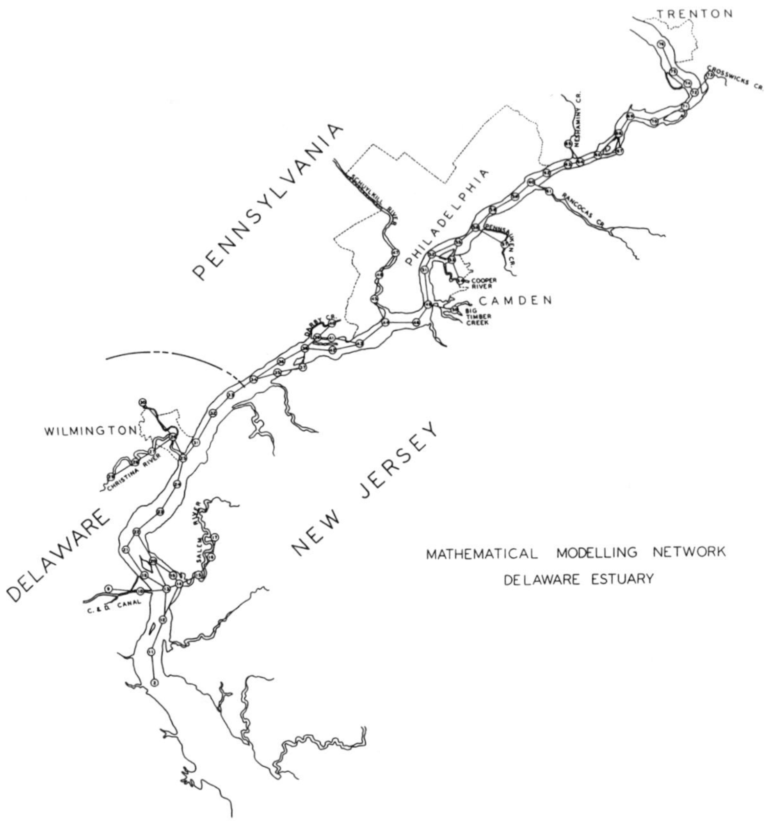

FIGURE 5. Link-nodes for Delaware Estuary model.

Hartman and Hammond[20] indicated that their relationship fits the available field data within 20% for wind speeds between 3 and 12 m/sec.

IV. BOD/DO CASE STUDIES

A. THE DELAWARE ESTUARY

The Delaware Estuary (Figure 5) is one of the largest estuaries in the U.S. From the fall line at Trenton, New Jersey, the estuary extends a distance of 85 mi to the Delaware Bay. The large metropolitan centers of Philadelphia, Camden, and Chester discharge a significant amount of municipal and industrial wastewater that has a tremendous impact on the water quality.

The first generation of water pollution control efforts, largely completed by 1960, resulted in secondary treatment levels at most treatment plants above Philadelphia. Primary treatment was considered adequate in the estuary below Philadelphia. While most areas built the required facilities, some facilities from the first-generation effort were not completed until the 1960s or 1970s.

In 1961, the Delaware River Basin Commission (DRBC) was established. The DRBC has broad water resources responsibilities, including water pollution control. The commission developed a cleanup program based on a 6-year $1.2 million Delaware Estuary Comprehensive Study (DECS) made by the U.S. Public Health Service. Nearly 100 municipalities and industries were found to be discharging harmful amounts of wastes into the river. The DRBC calculated the river's natural ability to assimilate wastes and established allocations for each city and industry. The objective of the DRBC wasteload allocation program and the corollary programs of Pennsylvania, New Jersey, Delaware, and the federal government was to upgrade the somewhat improved water quality of 1960 to more acceptable levels.

One of the early water quality models of the Delaware Estuary was developed under the Delaware Estuary Comprehensive Study (DECS). The modeling approach employed in that study was the application of the technique of finite segments to the steady-state and time-dependent (tidally averaged) conservation of mass equations.[21-23] In the modeling studies, the Delaware Estuary was divided into 30 segments, each of which was assigned a (midtide) volume and cross-sectional areas at the junctions. The DECS model was used in issuing wasteload allocations for the point source dischargers in 1968. In 1967, based on the DECS model, the commission adopted new, higher water quality standards and then in 1968 issued wasteload allocations to approximately 90 dischargers to the estuary. These required treatment levels were more stringent than secondary treatment as defined by the EPA.

In 1978, the U.S. EPA adopted the Dynamic Estuary Model (DEM) for the Delaware Estuary in a one-dimensional configuration.[24] The DEM consists of two separate but interrelated components: a hydraulic program dealing with water motion and a quality program dealing with mass transport and water quality. The DEM was originally developed during the mid-1960s while estuarine modeling was still in its infancy.[25] The DEM was innovative in considering a real-time computerized tidal solution of the hydrodynamic and water behavior of estuarine environments.

The water quality model consists of two separate but related components: the hydrodynamic module and water quality module. In a typical application, the hydrodynamic module is first run, yielding a description of flow patterns in the system. This information is summarized and serves as input to the water quality module which is applied next in order to define the interaction of physical, chemical, and biological processes as they affect the water quality parameters of interest.

The model can accommodate a range of time and space scales as may best suit the nature of the problem and the physical characteristics of the estuary. Computations of tidal flow and stage are conducted at time steps from 0.5 to 5.0 min and at spatial intervals from a few hundred to several thousand feet. In water quality calculations, the concentrations are computed on the same spatial scales as the hydrodynamic computations but are on expanded time scales of about 0.25 to 1.0 h. Thus, the model is dynamic in nature; it calculates real-time flows and concentrations. At the end, tidally averaged concentrations over a number of tidal cycles can also be computed for steady-state conditions. Five water quality constituents are simulated by the model: CBOD, organic nitrogen, ammonia, and nitrite/nitrate, and dissolved oxygen. The model kinetic structure is shown Figure 6.

The one-dimensional water quality model has been calibrated using a number of data sets, one of which is the July 1976 data.[26] That data was derived from a high water slack survey and a low water slack survey conducted from July 12, 1976 to July 23, 1976 with an average water temperature of 25°C. The freshwater flow at Trenton is 7700 *cfs* and the flow in the Schuylkill entering the Delaware is 1350 *cfs*. The wastewater loading rates during that period are listed in Table 3. Figure 7 presents the model results in conjunction with the field data during July 1976 for four water quality parameters: ammonia, nitrite/nitrate, $CBOD_5$, and dissolved oxygen. There is no data available on organic nitrogen concentrations and therefore their model results are not presented. In general, the model mimics the water quality trends in the Delaware Estuary very well. The model results presented in Figure 7 are quite similar to those obtained by Clark et al.[27] The dissolved oxygen sag occurred at about 35 mi downstream from Trenton.

In 1982, the EPA expanded the one-dimensional DEM model of the Delaware Estuary to a two-dimensional model to address the lateral variations of water quality concentrations in the estuary.[27] The model results were analyzed to quantify the relative importance of various components in the dissolved oxygen budget along the estuary. Figure 8 shows the results from a component analysis of dissolved oxygen deficits in the main channel of the Delaware Estuary with respect to upstream input, point sources (including municipal and industrial) CBOD deoxygenation, instream nitrification, sediment oxygen demand, and algal photosynthesis using the August 1975 low water slack survey data. The average freshwater flow at

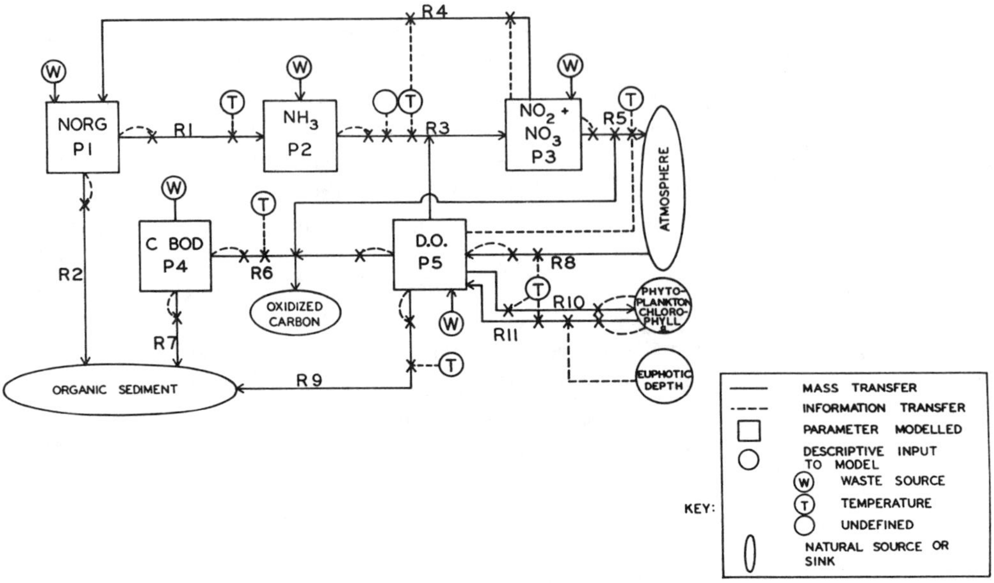

FIGURE 6. Water column kinetics of Delaware Estuary model.

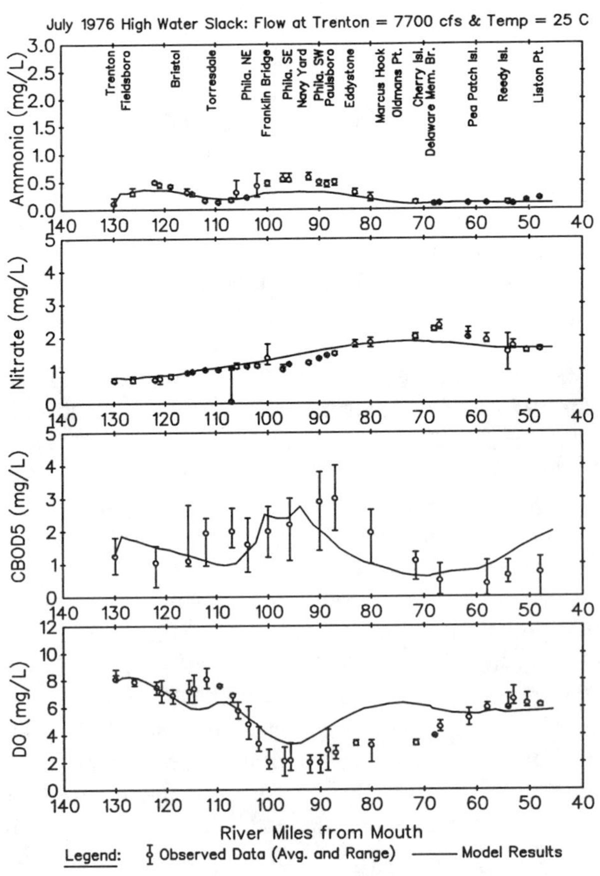

FIGURE 7. Delaware Estuary model results — July 1976.

TABLE 3
Point Source Waste Loads[a] to Delaware Estuary (July 1976)

Node no.	Name of discharge	Type of discharge	Flow[b] (mgd)	NH_3	NO_{23}^-	$CBOD_u$	DO
14	SALEMCTY	MUN	2.80	0.20	0.00	2.24	0.12
17	SALEM CR	TRIB	2.39	0.00	0.00	0.22	0.00
21	GETTYOIL	IND	8.00	2.49	0.06	3.00	0.20
22	AMOCO	IND	0.60	0.09	0.00	0.97	0.00
23	PENNSVLE	MUN	0.90	0.22	0.00	1.95	0.00
24	DPCHAMBR	IND	100.00	10.02	10.85	170.61	3.34
24	ICI 1	IND	2.60	0.00	0.04	1.35	0.09
24	ICI 7	IND	1.10	0.00	18.73	0.44	0.05
24	ICI 8	IND	0.10	0.00	0.00	0.08	0.01
24	ICI 13	IND	0.90	0.54	0.01	7.51	0.01
25	UPENSNCK	MUN	0.50	0.00	0.00	1.07	0.00
25	WLMINGTN	MUN	70.50	7.65	1.35	18.83	2.35
29	CHRSTINA	TRIB	148.39	0.28	1.88	11.15	4.95
30	BRANDYWN	TRIB	304.97	0.15	5.32	48.87	20.87
31	DPEDGMOR	IND	8.00	0.02	0.13	0.39	0.33
31	PENSBROV	MUN	0.30	0.00	0.00	0.13	0.00
33	PHOENIX	IND	11.00	0.23	0.18	0.51	0.00
33	ALLDCHEM	IND	24.40	0.31	0.41	2.66	0.41
33	OLDMANS	TRIB	31.16	0.04	0.32	1.77	1.30
34	MARCUSHK	MUN	0.60	0.08	0.01	1.38	0.00
34	SUNOIL 1	IND	74.00	1.79	1.73	28.66	1.85
34	FMC	IND	2.50	0.00	0.06	4.76	0.13
34	BP 201	IND	2.20	0.06	0.01	0.80	0.07
34	BP 101	IND	74.00	0.19	1.54	6.27	1.85
34	BP 002	IND	38.00	0.13	0.79	2.30	1.27
34	CHESTER	MUN	8.80	1.15	0.08	21.01	0.07
34	MONSANTO	IND	1.80	0.66	0.00	4.07	0.02
36	CHESTER	TRIB	25.81	0.06	1.36	1.16	1.72
36	SCOTT 2	IND	6.70	0.01	0.13	11.11	0.39
36	SCOTT 3	IND	7.50	0.01	0.13	9.08	0.44
36	SCOTT 4	IND	3.90	0.01	0.07	5.71	0.23
39	UCARBIDE	IND	2.60	0.30	0.08	0.70	0.05
40	CDCA	MUN	8.00	1.34	0.05	6.68	0.20
40	DRBYCRSA	MUN	11.00	1.99	0.17	5.88	0.28
40	MUKNPATS	MUN	5.00	0.50	0.24	1.67	0.11
40	DARBY CR	TRIB	18.06	0.18	0.33	1.12	0.45
42	PRPAUN	IND	15.00	0.63	2.75	5.63	0.75
42	HERCULES	IND	0.60	0.00	0.00	0.04	0.00
43	MOBILCP1	IND	13.30	0.00	0.67	1.29	0.56
43	MOBILNY2	IND	4.70	0.10	0.40	0.84	0.15
43	MOBILIW3	IND	4.30	1.04	0.00	3.96	0.00
43	GLOSTRCO	MUN	11.40	0.54	0.55	1.71	0.48
43	PAULSBRO	MUN	1.30	0.29	0.00	1.39	0.00
43	SHELL	IND	1.90	0.06	0.00	0.67	0.00
43	OLIN CHEM	IND	17.40	0.30	0.00	0.69	0.00
43	MANTUA	TRIB	7.10	0.30	0.04	0.56	0.00
44	PHILA SW	MUN	140.00	7.24	0.71	102.83	4.67
44	WOODBURY	MUN	1.90	0.02	0.00	2.71	0.00
44	NAT PARK	MUN	0.60	0.02	0.00	0.64	0.00
44	GULF OIL	IND	9.30	0.15	0.93	2.03	0.36
44	ARCO SPL	IND	3.50	.01	0.11	0.21	0.18
44	ARCO WPL	IND	0.10	0.00	0.00	0.01	0.00
44	ARCO NYD	IND	2.20	0.01	0.01	1.70	0.03
48	TEXACO	IND	4.40	0.00	0.17	0.53	0.18
48	GLOSTRCY	MUN	3.00	0.35	0.00	2.70	0.00

TABLE 3 (continued)
Point Source Waste Loads[a] to Delaware Estuary (July 1976)

Node no.	Name of discharge	Type of discharge	Flow[b] (mgd)	NH_3	NO_{23}^-	$CBOD_u$	DO
48	NJ ZINC	IND	4.00	0.02	0.00	0.34	0.07
48	BIGTIMBR	TRIB	45.16	0.32	0.64	3.77	2.11
47	SCHUKILL	TRIB	1350	0.29	16.36	37.80	58.16
49	PHILA SE	MUN	131.00	2.30	1.53	238.37	4.37
49	MCAND & FB	IND	1.30	0.00	0.00	5.56	0.00
49	CAMDEN M	MUN	26.00	2.17	0.00	65.10	0.22
49	HARSHOW	IND	0.60	0.00	0.00	1.28	0.00
49	GAF	IND	11.00	0.00	0.00	3.06	0.00
50	NEWTON	TRIB	3.87	0.03	0.04	0.11	0.09
51	AMSTAR 1	IND	1.90	0.01	0.03	0.10	0.06
51	AMSTAR 3	IND	23.6	0.08	0.32	1.30	0.69
52	NATSUGAR	IND	15.80	0.04	0.09	1.32	0.26
54	COOPER	TRIB	28.39	0.71	0.31	2.84	1.04
54	CAMDEN N	MUN	3.20	0.67	0.00	5.77	0.00
55	PHILA NE	MUN	172.00	14.21	0.72	272.77	3.73
55	PENSAUKN	MUN	4.20	0.60	0.00	11.36	0.00
55	GEORGPAC	IND	0.80	0.00	0.00	6.72	0.01
57	PENSAUKN	TRIB	11.61	0.19	0.16	1.24	0.29
58	PALMYRA	MUN	0.40	0.11	0.01	0.32	0.01
59	PENNYPAK	TRIB	3.23	0.00	0.09	0.05	0.28
59	CINAMNSN	MUN	2.00	0.27	0.00	1.00	0.05
61	WLINGBRO	MUN	1.90	0.00	0.00	1.22	0.00
64	RANCOCAS	TRIB	154.84	0.26	1.81	9.05	9.05
64	BRLINGTN	MUN	1.20	0.04	0.00	3.00	0.00
64	TENNECO	IND	1.30	0.20	0.03	0.75	0.04
65	NESHAMNY	TRIB	100.26	0.10	2.51	2.85	6.95
65	FALLSTWP	MUN	2.60	0.05	0.24	0.42	0.00
66	BRSTLTWP	MUN	1.70	0.21	0.01	0.57	0.03
66	BRSTLBRO	MUN	1.70	0.11	0.04	1.19	0.10
66	ROHM & HAS	IND	1.00	0.00	0.00	0.37	0.04
66	OTTERCRK	TRIB	3.23	0.00	0.00	0.66	0.00
66	ASSISCNK	TRIB	3.23	0.00	0.00	0.66	0.00
69	FLORENCE	MUN	0.60	0.06	0.00	0.64	0.03
69	LWRBUCKS	MUN	7.90	1.32	0.03	2.37	0.24
69	BRIPARCH	IND	3.20	0.01	0.00	0.68	0.00
69	MARTINS	TRIB	5.48	0.00	0.00	0.33	0.00
71	USSRODMN	IND	7.00	0.16	0.07	0.00	0.35
71	USSTRMTP	IND	61.3	1.18	0.77	2.05	2.56
72	BORDENTN	MUN	1.00	0.12	0.0	0.81	0.04
73	HAMILTON	MUN	8.60	1.79	0.22	2.58	0.29
73	CROSWICK	TRIB	40.65	0.05	0.42	2.04	2.37
75	TRENTON	MUN	19.0	7.14	0.00	28.55	0.16
76	MORRISVL	MUN	3.7	0.97	0.06	1.11	0.23
76	ASSNPINK	TRIB	65.81	0.37	1.32	3.30	3.95

[a] All loads in 1000 lb/d.
[b] Tributary flows in *cfs*.

Trenton during this survey is 7880 *cfs* and the average temperature in the estuary is 27°C. The calculated profiles in Figure 8 show that the oxidation of wastewaters from POTWs and industrial facilities (in CBOD) is the most significant source of dissolved oxygen deficit in the Delaware Estuary. At the point of maximum deficit, the oxidation of CBOD from these dischargers is responsible for about 5 mg/l of dissolved oxygen deficit in the water column.

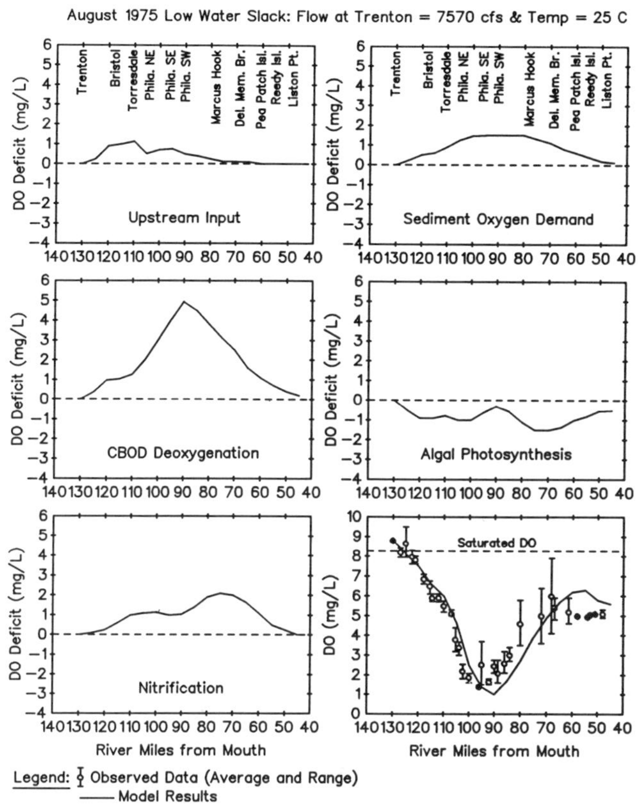

FIGURE 8. Dissolved oxygen component analysis for the Delaware Estuary model.

Note that the downstream boundary condition has a dissolved oxygen deficit about 2 mg/l (not shown in Figure 8), preventing a complete recovery of the dissolved oxygen concentration in the water column at the downstream end. The component analysis allows the identification and quantification of important kinetic process(es) in contributing to the dissolved oxygen deficit in the water column.

The DEM model of the Delaware Estuary had a number of capabilities that represented significant improvements at the time of its development. On the other hand, it also has several important limitations. For example, nutrient uptake as a result of algae growth is considered only by the transformation of nitrate into organic nitrogen. The uptake of ammonia, as well as a preference term for distinguishing between ammonia and nitrate, is not considered. The complicated interaction between dissolved oxygen, algae, photosynthesis, respiration, and light is not modeled at all, but instead, is reproduced through the judicious selection of constant P and R rates and euphotic depths.

B. THE JAMES ESTUARY

The James River system is one of the most important water resources in the Commonwealth of Virginia (Figure 9). Being the largest river in the state, the James River extends more than 400 mi from its mouth at Chesapeake Bay to its headwaters near the West Virginia state line. The river is a recognized asset to the surrounding residential and metropolitan areas, providing recreational opportunities, such as boating and fishing. It is also an asset to commerce and industry, serving as an important water supply, and as such, are a catalyst for economic growth.

The estuarine system starts near Richmond where the fall line is located and is approximately 100 mi from the mouth of the river. The primary water quality concern in the estuarine system

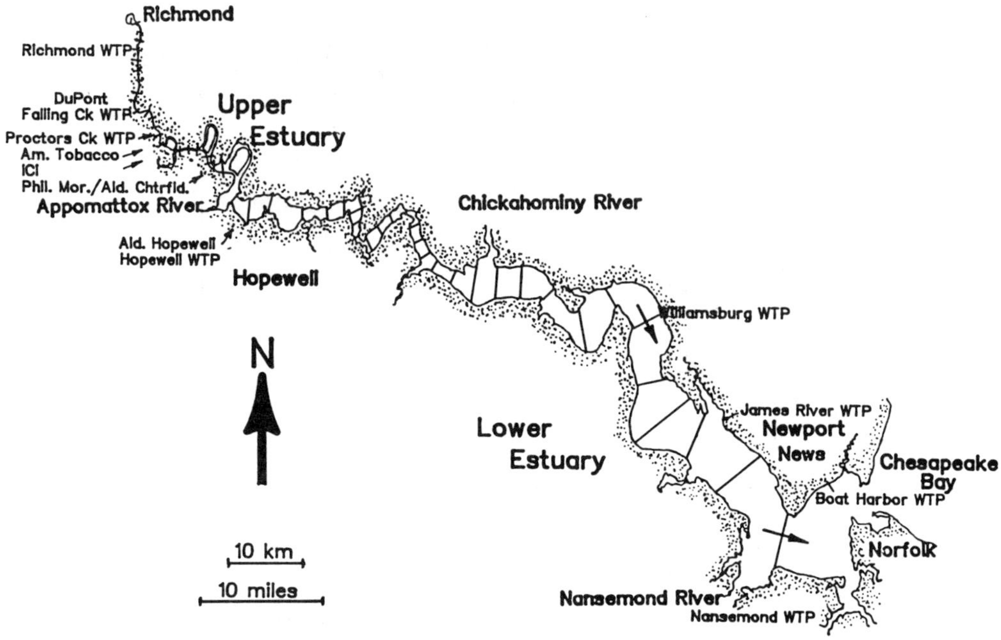

FIGURE 9. James River Estuary.

has been dissolved oxygen and increased nutrient loads. The dissolved oxygen is affected by the carbon and nitrogen components of the wastewater effluents.

Water supply and wastewater treatment facilities have been developing at a rate commensurate with growth in the James River basin over the past few decades. As a result, the James River, including the Appomattox River, has received increased quantities of treated effluent from both municipal and industrial sources. The Virginia State Water Control Board (SWCB) realized the necessity of planning for waste treatment requirements many years ago. Recognizing that proper planning must be implemented on a regional basis to protect the river system from impairment of its numerous desirable uses, SWCB entered into an agreement with the U.S. EPA in 1971, under Section 3(c) of the Federal Water Pollution Control Act of 1965, to study the James River. A principal outcome of this effort, completed in 1974, was the development of a James River ecosystem model by the Virginia Institute of Marine Science (VIMS). This model was later modified and the revised model, called JMSRV, was used by the SWCB staff to develop wasteload allocations, i.e., the *Upper James River Estuary Wasteload Allocation Plan*, in 1982. The model is able to simulate eight water quality constituents: $CBOD_u$, organic nitrogen, NH_3, NO_2/NO_3, organic phosphorus, ortho-phosphorus, chlorophyll *a*, and dissolved oxygen. The kinetic diagram relating these constituents is shown in Figure 10.

In 1986, the JMSRV model was expanded to include the Appomattox River.[28] Subsequently, Lung[29] modified the model to cover the entire estuary in 1991. The expanded model has been calibrated using several data sets and was then used to assess the water quality improvement following the treatment upgrade from primary to secondary at POTWs.[29] By 1980s, most of POTWs in the James River basin had achieved secondary treatment levels and beyond. Their flows and wasteloads in July 1983 are listed in Table 4.

For comparison purpose, Table 5 shows the concentrations in primary and secondary effluents used in the modeling analysis.[29] Although the major improvement of secondary treatment over primary treatment is the reduction of carbon (CBOD) contents, reduction of

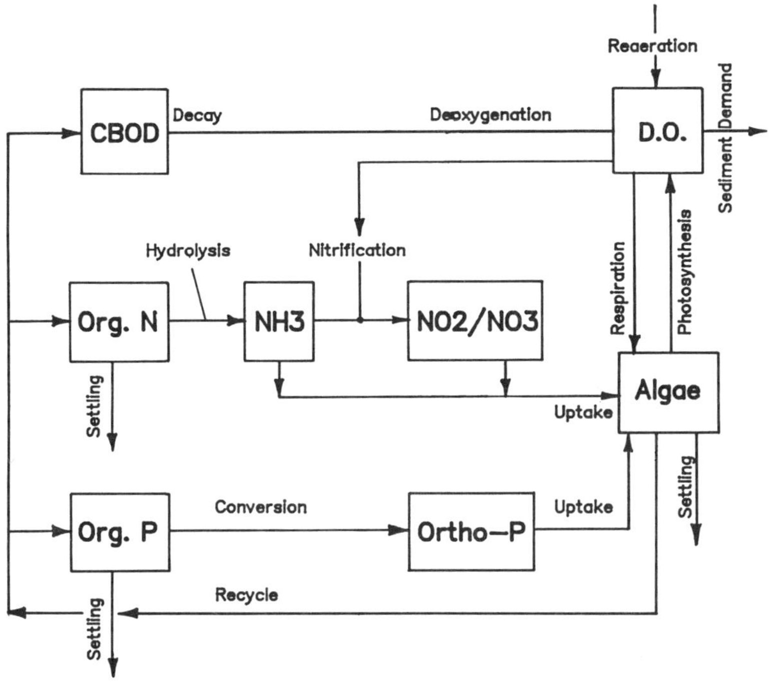

FIGURE 10. Segmentation and water column kinetics of James Estuary model.

ammonia via nitrification, to a certain extent, has been reported (Hall and Foxen 1983) at some activated sludge plants and is reflected in Table 5. In addition, the biological oxidation process also removes a certain amount of total phosphorus in the wastewater. Note the small increase of nitrite/nitrate due to partial nitrification in the biological oxidation process. For the James Estuary, instream nitrification in the river reach from Richmond to Hopewell was substantiated by a field study of the nitrifying bacteria.[28,29]

Model simulation scenarios for assessing the water quality improvement are summarized in Table 6. The freshwater flow in the James River at Richmond is 631 *cfs* and the freshwater flow in the Appomattox River near Petersburg is 65 *cfs*. The water temperature used in the simulations is 28°C.

Model simulation results are presented in Figure 11. The most significant result is the anoxic condition below Richmond in a reach about six miles long. Again, it should be pointed out that the actual dissolved oxygen levels under the 7Q10 condition and with all POTWs under primary treatment should be better than the result shown in Figure 11 simply because the wastewater flows were lower 20 years ago when the plants had primary treatment. Nevertheless, keeping the wastewater flow rates the same among all three simulation scenarios would provide a good perspective of the water quality improvement due to treatment upgrades. Also shown in Figure 11 are high concentrations of $CBOD_u$, organic nitrogen, ammonia, and nitrite/nitrate immediately below Richmond. Another peaks in $CBOD_u$ and organic nitrogen near the Hopewell area are associated with the algal biomass in that area.

Because of treatment upgrades, the $CBOD_u$ and nutrient concentrations below Richmond are much lower than those in scenario #1 with the exception of nitrate. Note that the secondary effluent has a higher nitrate concentration than the primary effluent. Dissolved oxygen levels improve significantly with the removal of the anoxic condition in the river although the sag concentration is still below 2 mg/l. Under the 1983 loading rates, the dissolved oxygen profile shows a much improved condition in the estuary with a DO sag about 4.8 mg/l.

TABLE 4
Major Point Source Loads to the James River Estuary in July 1983

Discharger	River (mi)[a]	Flow (mgd)	$CBOD_u$ (lb/d)	Org. N (lb/d)	Ammonia (lb/d)	Nitrate (lb/d)	Total P (lb/d)	Org. P (lb/d)	Ortho-P (lb/d)
Richmond	97.8	54.9	5642	1281	3216	1379	2314	144	2170
DuPont	92.7	8.4	427	217	0	63	12	6	6
Falling Creek	92.2	7.2	1067	398	328	311	461	109	351
Proctors Creek	86.9	2.7	312	312	45	36	156	64	91
Reynolds Metals	86.9	0.08	3	0	0	0	0	0	0
American Tobacco	81.5	0.75	16	60	14	3	40	17	22
ICI	80.6	0.07	17	8	0	4	1	1	0
Philip Morris	79.8	1.7	485	26	8	351	140	52	88
Allied-Chester	78.5	42.8	3859	46	3	35	0	0	0
Allied-Hopewell	77.2	161.6	16502	1163	1055	1514	60	47	13
Hopewell	76.1	22.9	19347	5046	6989	429	322	119	293
Williamsburg	38.0	9.2	306	176	54	8	160	40	120
James River	24.9	12.5	938	189	905	308	775	1	774
Boat Harbor	15.5	16.1	1611	188	2818	13	792	2	790
Nansemond	7.3	6.8	719	102	519	169	279	10	269
Army Base	6.5	12.1	1009	51	2017	10	450	0	459
Lambert's Points[b]		20.2	18400	921	3201	17	385	253	132
Petersburg[c]		12.7	1940	144	1875	856	250	50	200

[a] Distance from the Chesapeake Bay.
[b] Discharged into the Elizabeth River.
[c] Discharged into the Appomattox River and 10.8 miles from the James River.

TABLE 5
Effluent Concentrations in Primary and Secondary Effluents

Effluent	BOD_5 (mg/l)	Org. N (mg/l)	Ammonia (mg/l)	Nitrate (mg/l)	Total N (mg/l)	Org. P (mg/l)	Ortho-P (mg/l)	Total P (mg/l)	DO (sat.)
Primary	115	9.7	16.4	0.30	26.4	2.7	6.0	8.7	50%
Secondary	30	8.6	10.7	3.90	23.2	0.3	3.7	4.0	50%

TABLE 6
Model Simulation Scenarios

Scenario no.	Waste load	K_d (d^{-1})	$CBOD_u/CBOD_5$
1	(Table 5)	0.20	1.5
2	(Table 5)	0.16	2.84
3	(Table 4)	0.10	[a]

[a] Actual values.

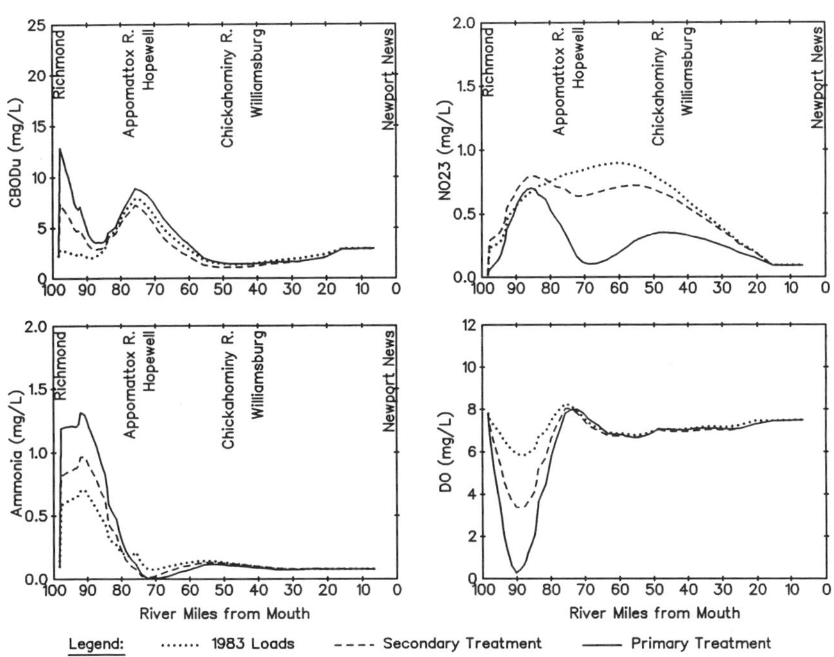

FIGURE 11. James Estuary model simulation results following treatment upgrades.

V. EUTROPHICATION KINETICS

Figure 12 presents the major kinetic processes for eutrophication modeling. Processes 1 to 7 are already described earlier in Figure 1 and included in eutrophication analysis. Other processes related to algal growth and nutrient recycling are:

- Sorption, desorption of inorganic material
- Settling and deposition of phytoplankton
- Uptake of nutrients and growth of phytoplankton

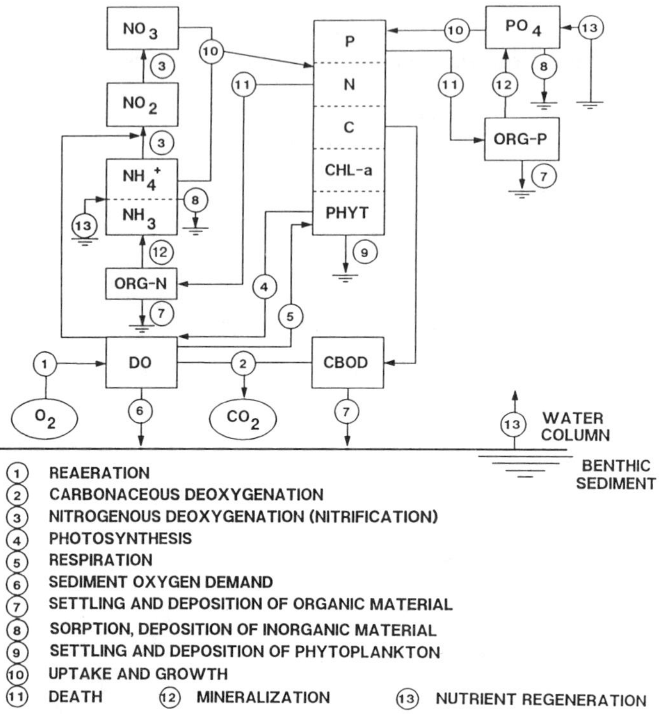

FIGURE 12. Eutrophication kinetic processes in water column.

- Death of phytoplankton
- Mineralization of organic nutrients
- Nutrient regeneration from the sediment

A. TEMPERATURE EFFECT ON PHYTOPLANKTON GROWTH

The growth coefficient is directly related to temperature in moderate climates. Auer and Canale[30] and Canale and Vogel[31] summarized data from phytoplankton growth experiments conducted at various temperatures. These results, plotted as the solid and dashed lines in Figure 13a, illustrate the different temperature optimums for different phyla of phytoplankton and also the differences in the way temperature influences the growth rate. Essentially, Figure 13a can be utilized to characterize the growth rates as a function of temperature for diatom, green, and blue-green species. Eppley[32] summarized algal growth data from a variety of sources to define an expression that represent optimum growth as a function of temperature as follows:

$$G_T = G_{max} \theta^{(T-20)} \tag{4.11}$$

where

G_T = temperature adjusted growth rate (d^{-1})
G_{max} = maximum growth rate at 20°C (d^{-1})
θ = constant for temperature adjustment
T = temperature (°C)

Both G_T and G_{max} are specific growth rates under optimum light and nutrient conditions. Reported ranges for G_{max} and q are:

G_{max} = 1 to 3 d^{-1} at 20°C
θ = 1.01 to 1.18

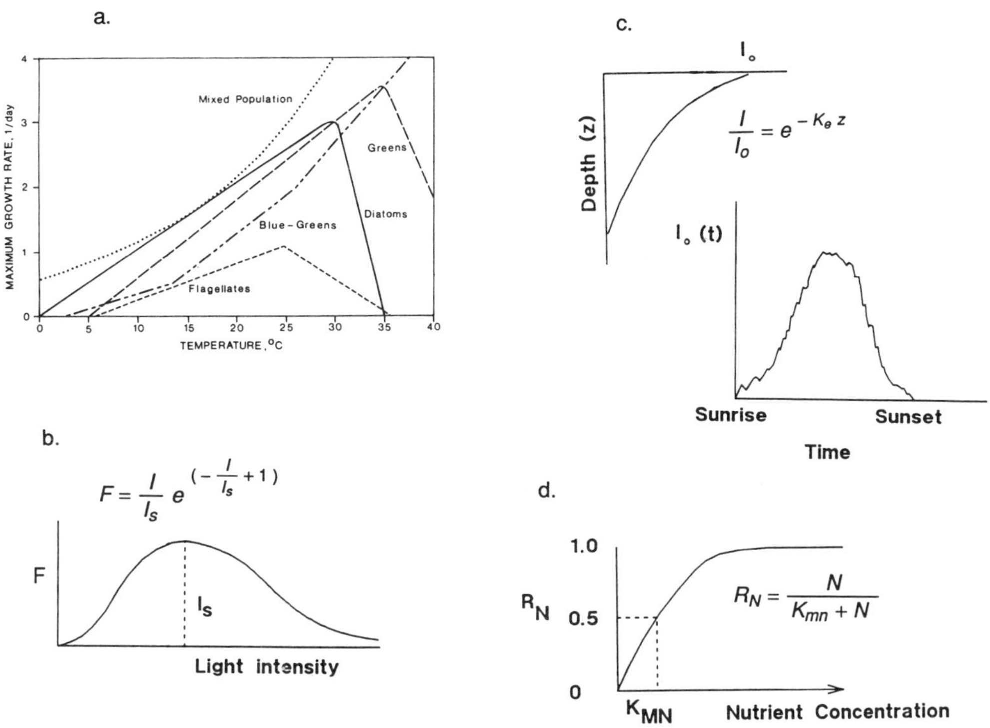

FIGURE 13. Effects of temperature, light, and nutrients on phytoplankton growth.

As a first approximation, $G_{max} = 1.8\ d^{-1}$ and $\theta = 1.066$ may be used.[1] Thus,

$$G_T = 1.8(1.066)^{(T-20)} \tag{4.12}$$

Equation 4.12 is shown as the dotted curve in Figure 13a. It can be viewed as an envelope representing the maximum growth rate at any temperature, under optimum light and nutrient conditions.

B. LIGHT EFFECT ON PHYTOPLANKTON GROWTH

The growth rate of phytoplankton is also dependent on the light intensity up to a saturating condition, greater than which it decreases with light (see Figure 13b). The growth rate at saturating light condition can be expected to be species dependent as in Figure 13b. Because light energy available to phytoplankton varies so much with depth and time of day, an appropriate expression of light availability for use in analyses should account for these changes. A depth and time averaged effect of available light energy on phytoplankton growth rate can be obtained, by integrating the light intensities relationships over depth and time.[33] This reduces to

$$r_l = \frac{2.718f}{K_e HT}\left(e^{-\frac{I_f}{I_s}e^{-K_e H}} - e^{-\frac{I_f}{I_s}} \right) \tag{4.13}$$

where

r_l = light limitation factor
f = photoperiod — daylight fraction of averaging period
T = averaging period (1.0 d)
K_e = light extinction coefficient (m^{-1})

H = average depth of segment (m)
I_a = average of incident light on water surface over a 24-h day (ly/d)
I_f = average of incident light over photoperiod (=I_a/f)
I_s = saturated light intensity (ly/d) (see Figure 13c)

Solar radiation is measured routinely at selected weather stations in the U.S. It is usually reported as langleys (ly), which is a measure of the total radiation of all wave lengths that reaches the surface of the earth during a 24-h period.

The light reduction factor, r_l, interpreted as the percentage of the optimum growth rate, is sensitive to the product of $K_e H$ which appears as the denominator in Equation 4.13. It is seen that shallow and clear waters yield high r_l values and offer a favorable condition for algal growth when compared with turbid and deep waters. $K_e H$, a dimensionless number, is also referred to as the light extinction factor.

Extinction coefficients can be determined directly using light intensity measurements from the field. Light attenuation with depth is approximated by the following equation:

$$I = I_o e^{-K_e z} \tag{4.14}$$

That is, the slope of $ln\ (I/I_o)$ vs. depth, z, provides an estimate of K_e. In lieu of the direct measurements of light intensity at various depths, K_e may be determined by the following empirical equation:

$$K_e = \frac{1.6}{\text{Secchi Depth}} \tag{4.15}$$

It should be pointed out that the correlation between Secchi depth and light extinction coefficient is notoriously poor in many waters; factors may range from 1.5 to 5.0. Thus, field determination of K_e is recommended.

Di Toro[34] has provided a theoretical and empirical basis for estimating the extinction coefficient as a function of nonvolatile suspended solids, detritus, and phytoplankton chlorophyll:

$$K_e = 0.052\ NVSS + 0.174\ VSS + 0.031\ Chl \tag{4.16}$$

where

K_e = light extinction coefficient (m^{-1})
$NVSS$ = nonvolatile suspended solids concentration (mg/l)
VSS = detritus (nonliving organic carbon) concentration (mg/l)
Chl = chlorophyll concentration (mg/l)

The nonvolatile suspended solids (the inorganic particulates) both absorb and scatter the light whereas the organic detritus and phytoplankton chlorophyll mainly absorb the light. Di Toro[34] has shown that Equation 4.16 applies to K_e values of about less than 5.0 m^{-1}.

C. NUTRIENT EFFECT ON PHYTOPLANKTON GROWTH

The phytoplankton growth rate is also a function of nutrient concentrations up to a saturating condition, greater than which it remains constant with nutrient concentration (Figure 13d). At zero nutrient concentration, there is no growth. As the nutrient level is increased, growth begins. However, as nutrient levels continue to increase, the effect on the growth rate of the phytoplankton is reduced and asymptotically approaches unity. Such a relationship is described by a

Michaelis-Menton formulation whose significant parameter is that concentration at which the growth rate is equal to one half of that at the saturated concentration:

$$r_n = \frac{N}{K_m + N}$$

where

r_n = nutrient reduction factor for algal growth
N = the nutrient concentration (mg/l)
K_m = half saturation (Michaelis) constant (mg/l)

The Michaelis constant is a function of algal species. Their values usually range from 5 to 25 μg/l for nitrogen and from 1 to 5 μg/l, depending on the species. With more than one nutrient present (i.e., nitrogen and phosphorus), the nutrient effect is given by

$$r_n = \min\left(\frac{DIN}{K_{mn} + DIN}; \frac{DIP}{K_{mp} + DIP} \cdots\right) \tag{4.17}$$

where

r_n = nutrient reduction factor
DIN = inorganic nitrogen concentration (sum of ammonia, nitrite, and nitrate)
DIP = dissolved inorganic phosphorus concentration
K_{mn} = Michaelis-Menton constant for nitrogen (mg/l)
K_{mp} = Michaelis-Menton constant for phosphorus (mg/l)

The individual nutrient limitation values are computed and minimum value is chosen for the overall nutrient limitation for algal growth.

Figure 14 shows the Michaelis-Menton formulation in a slightly different format. In this figure, K_{mn} = 25 μg/l and K_{mp} = 1 μg/l are used. For an estuary with a DIN concentration of 100 μg/l, this corresponds to a 20% reduction in the growth rate (r_n = 0.8). In order for phosphorus to become the limiting nutrient in the system, dissolved inorganic phosphorus must reach a level of 4 μg/l or less. It should also be pointed out that if upstream nitrogen controls were instituted such that DIN was reduced to 60 μg/l for the same reach of the stream then a further reduction in DIP to 2.5 μg/l would be required to keep phosphorus as the limiting nutrient. In other words, as the water column concentrations of DIP begin to approach growth limiting levels due to continued reduction in point source phosphorus effluents, any nitrogen control strategies that might be instituted would require additional levels of phosphorus removal in order to keep phosphorus as the limiting nutrient.

The full expression for algal growth can be synthesized from Equations 4.12, 4.13, and 4.17 as follows:

$$G_p = G_{\max} 1.066^{(T-20)} \left[\frac{2.718 f}{K_e H T}(\alpha_1 - \alpha_2)\right] \min\left(\frac{DIN}{K_{mn} + DIN}; \frac{DIP}{K_{mp} + DIP}\right) \tag{4.18}$$

where

$$\alpha_1 = e^{-\frac{I_f}{I_s} e^{-K_e H}}$$

$$\alpha_2 = e^{-\frac{I_f}{I_s}}$$

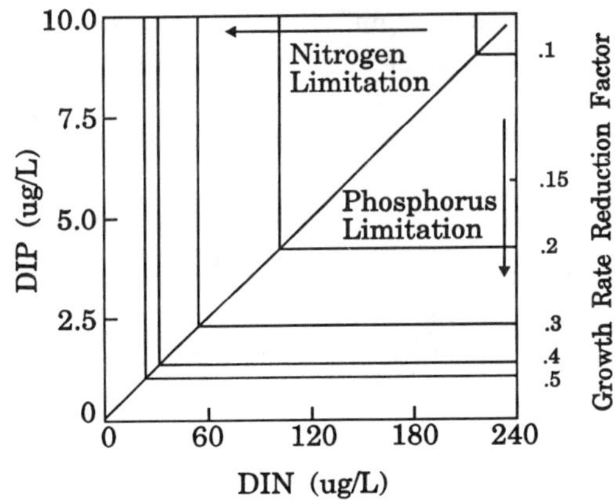

FIGURE 14. Michaelis-Menton relationship for nutrient limitation.

D. ALGAL DEATH

Decreases in algal concentrations are brought about by two processes:

- Algal respiration
- Death

Algal respiration is caused by endogenous respiration in which algal biomass is oxidized to generate CO_2. Algal death includes grazing by zooplankton (for diatoms and greens only) and cell destruction through bacterial attack, disease, physical damage, the natural aging process, or other mechanisms. The distinction between phytoplankton reductions through death and reductions through respiration, grazing by zooplankton, or settling is that upon death, all the carbon, nitrogen, and phosphorus contained in the algal biomass is returned to the carbonaceous BOD (CBOD) and organic nitrogen and phosphorus pools, respectively. During respiration, carbon is given off as CO_2 rather than CBOD; through grazing, only portion of the organic contents of the algal cells is returned to the respective organic pools (the remaining portion is lost from the balance as zooplankton mass).

The algal reduction rate can be expressed as:

$$D_p = D_{p1}(T) + D_z + D_d \tag{4.19}$$

where

$D_{p1}(T)$ = temperature dependent endogenous respiration rate (d^{-1})
D_z = phytoplankton loss rate to grazing (d^{-1})
D_d = non-predatory mortality rate (d^{-1})

The phytoplankton death rate, D_z, is a function of zooplankton population and zooplankton grazing rate.

E. ALGAL SETTLING

Phytoplankton are lost from the water column through settling. In a vertically mixed water column, the net settling rate (i.e., settling to the bottom less resuspension from the bottom) is expressed as:

$$S = \frac{V_s}{H}$$

where

S = phytoplankton settling rate (d^{-1})
V_s = phytoplankton settling velocity (m/d)
H = average depth (m)

Through settling, none of the organic cell material is returned to the organic pools.

F. NITROGEN COMPONENTS

The major components of the nitrogen system are detrital organic nitrogen, ammonia, nitrite, and nitrate. In natural waters, there is a stepwise transformation from organic nitrogen to ammonia, nitrite and nitrate, yielding nutrients for phytoplankton growth. The kinetics of the transformations are temperature dependent.

During algal respiration and death, a fraction of the cellular nitrogen is returned to the nitrogen pool in the form of ammonia nitrogen. The remaining fraction is recycled to the organic nitrogen pool. Organic nitrogen undergoes a bacterial decomposition whose end product is ammonia nitrogen. Ammonia, in the presence of nitrifying bacteria and oxygen, is oxidized to nitrate nitrogen (nitrification). Denitrification by bottom sediments may be a major loss mechanism in some systems.

Both ammonia and nitrate are available for uptake and use in algal growth, however for physiological reasons the preferred form of nitrogen is ammonia. The ammonia preference term is characterized in Figure 15. For a given concentration of ammonia, as the available nitrate increases above approximately the Michaelis limitation the preference for ammonia reaches a plateau. Also as the concentration of available ammonia increases the plateau levels off at values closer to unity, i.e., total preference for ammonia.

Another important factor in modeling the fate and transport of nitrogen is the toxicity of un-ionized ammonia (NH_3). Ammonia speciation is pH- and temperature dependent. In practice, if pH in the water column is not expected to vary, it may be justified to simulate the nitrogen cycle in order to predict total ammonia concentrations.[35]

G. PHOSPHORUS COMPONENTS

In many stream water quality models, phosphorus is accounted for in two forms: dissolved and particulate. A fraction of the phosphorus released during phytoplankton respiration and death is in the inorganic form and readily available for uptake by other viable algal cells. The remaining fraction released is in the organic form and must undergo a mineralization or bacterial decomposition into inorganic phosphorus before utilization by phytoplankton.

There is an adsorption-desorption interaction between dissolved inorganic phosphorus, and suspended particulate matter in the water column. The subsequent settling of the suspended solids together with sorbed inorganic phosphorus can act as a significant loss mechanism in the water column and is a source of phosphorus to the sediment. Compared with the reaction rates for the algal and biological kinetics which are in the order of days, the adsorption-desorption rates are much faster, permitting an instantaneous equilibrium assumption for the calculation.

$$f_{DIP} = \frac{1}{1 + K_{PIP} SS}$$

where

f_{DIP} = fraction inorganic phosphorus dissolved
SS = suspended solids concentration (kg/l)
K_{PIP} = partition coefficient (l/kg)

Subsequent settling of the solids and sorbed phosphorus can provide a significant loss mechanism of phosphorus from the water column to the benthos. A wide range of partition

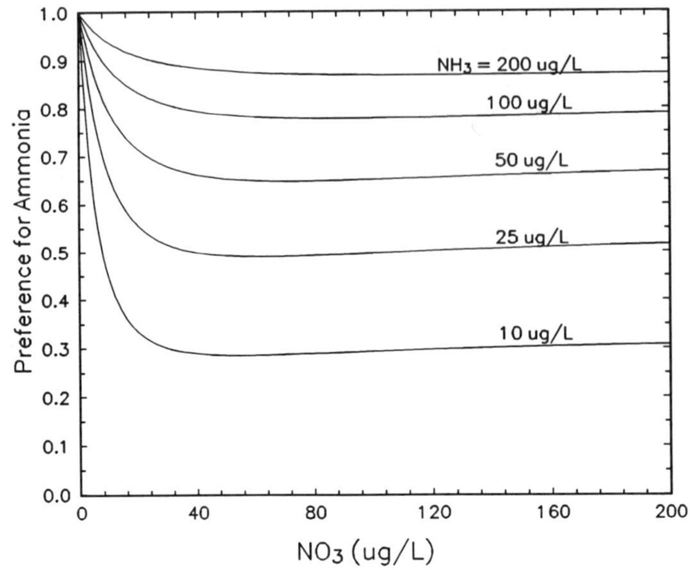

FIGURE 15. Ammonia preference by phytoplankton.

coefficients have been found in the literature for phosphorus. Process based functions that accurately calculate the phosphorus partition coefficient would improve prediction of this important variable significantly.[35] In model formulations, the concentrations of dissolved and particulate phosphorus need to be repartitioned at every time step.[19]

H. SEDIMENT NUTRIENT RELEASE

Sediment processes may have profound effects on dissolved oxygen and nutrients in some systems. Nutrient release occurs as a result of a gradient in nutrient concentration between the overlying water and the nutrient in the interstitial water of the sediment. In some systems, the impact of sediment nutrient release can be significant and result in continuing eutrophication problems even after point sources have substantially reduced through control measures. In addition, denitrification by sediments is often a major loss mechanism for nitrogen (Figure 12).

Although often of critical importance, the predictive capability of most presently available models of sediment interactions is limited.[35] Additional discussions on sediment interactions modeling is provided in Chapter 6.

VI. EUTROPHICATION CASE STUDIES

Estuarine eutrophication modeling started over 20 years ago with the first published work by Di Toro et al.[36] on the Sacramento-San Joaquin Delta in California. The progress in modeling of the freshwater systems such as lakes in the 1970s contributed to significant advancement in water column kinetics. The most noted early examples are the Great Lakes work by Thomann who worked on Lake Ontario and Di Toro who modeled Lake Erie. A more recent study on the Saginaw Bay was conducted by Bierman and Doland.[37] In recent years, the concern of the Chesapeake Bay in the east coast of U.S. continues the advancement of modeling technology on eutrophication modeling. In the 1970s, eutrophication modeling of estuaries focused on seasonal steady-state conditions while in the 1980s, time-variable modeling for an entire year was explored with reasonable success. At the time being, the forefronts of estuarine eutrophication modeling are the integration of hydrodynamic and water quality models and the development and incorporation of sediment layers to investigate the long-term recovery of an estuarine system.

A. SEASONAL STEADY-STATE PHYTOPLANKTON MODELING

Modeling phytoplankton growth in an estuarine system is complicated by many factors. A number of biochemical, biological, and chemical processes interact and their reaction rates vary with time. In addition, the freshwater flow and associated circulation and mass transport patterns also are functions of time, with time scales ranging from minutes to weeks. External inputs (forcing functions) and other parameters further complicated the time-variable nature of phytoplankton population in an estuarine system. In general, phytoplankton modeling analyses for estuarine systems may employ time-variable calculations in either real-time (intratidal) or tidally-averaged time-variable mode. Time-variable phytoplankton modeling is resource demanding and computationally intensive, often resulting in complex models. In addition, the amount of data required to calibrate and verify a real-time model is formidable. Data on hour-to-hour changes in tidal stage, velocity, freshwater inflow, and nutrient loads must be available for model calibration. For example, to calibrate a real-time model of the Manasquan Estuary, intensive sampling was conducted every three hours for 30 d in the tidal portion of the estuary and for ten days in the freshwater portion of the estuary.[38,39] Yet, depending on the nature and context of the water quality problem(s), most real-time modeling results are time-averaged for subsequent use in wasteload allocations. A practical example is the Delaware Estuary modeling study (see Section IV.A) using the two-dimensional Dynamic Estuary Model (DEM). The Delaware DEM results were tidally-averaged for use in the water quality analysis.[27] While the real-time modeling analysis may not be completely suitable for phytoplankton modeling, it is best used to simulate highly transient, localized events, such as combined sewer overflows or nonpoint source flows because the results provide a more realistic representation of hour-to-hour pollutant transport.

In certain situations, tidally averaged time-variable models are used, recognizing that minute-to-minute or hour-to-hour variation is not appropriate for algal growth dynamics. It is argued that all of the inputs that are relevant to the growth dynamics cannot be specified on so fine a time scale and the phytoplankton population does not significantly vary from hour to hour.[40] Thus, tidally averaged time-variable calculations are appropriate when seasonal water quality changes are more important than diurnal fluctuations. Intensive water quality surveys must be conducted throughout several cycles for each flow condition and the results averaged to obtain representative data. Intensive surveys should be performed throughout the year to reflect seasonal variability in tides and freshwater flows. However, such water quality data is not always available. Instead, many estuaries are sampled during slacks of a tidal cycle. Recent modeling studies using slack water survey data include the Patuxent River Estuary[41] and the Potomac Estuary.[19] These models simulate time-variable water quality conditions throughout the entire year on a tidally averaged or slack water approximation basis. The computational effort, although much less involved than the real-time intratidal calculations, is still intensive. In contemplating the time-variable vs. steady-state modeling, the following questions are often raised:

- Does the benefit from the time-variable analysis outweigh the effort?
- Under what conditions, seasonal steady-state approximations can be justified?
- Can additional approximation be made without compromising the model results?

In many estuarine systems, phytoplankton growth is a seasonal event. Thus, approximations of the phytoplankton-nutrient dynamics on a seasonal steady-state basis can be made. The approximation is particularly valid for an estuarine system under a summer low, steady flow condition. The advantages of a seasonal steady-state analysis include a modest data requirement still providing a significant amount of insight into the system.

One of the factors in considering steady-state approximations is the time for the systems to reach a steady-state (i.e., an equilibrium between the phytoplankton-nutrient dynamics in the water column and the external loading or forcing functions). Usually, freshwater flow, mass transport pattern, kinetics, and external input have significant bearing on the time to reach a steady-state.

The hydrological condition in temperate climates is characterized by high runoff and stream flows in the spring months followed by steady and reduced flows in the summer. The response time or the time to reach steady state is inversely proportional to the freshwater flow through the estuary. Therefore, the time to reach steady-state is shorter in spring than in summer. Based on the flow consideration, it can be stated that seasonal steady-state approximations to phytoplankton-nutrient dynamics would be more suitable for the spring months than the summer months. However, other factors, such as kinetics, play an important role in modifying the time to steady-state as discussed in the next paragraph.

Consider a simple case: a completely mixed water system in which residence time (hydraulic retention time) is dependent upon the flow rate through the system. The time to steady-state is approximately two to three times the residence time of the system under constant external loads. Further, settling of phytoplankton biomass tends to reduce the time to steady-state. That is, the loss of biomass due to settling would shorten the system response time. The amount of reduction depends on the relative magnitudes of flow and kinetic reaction rates (e.g., phytoplankton settling rate). The controlling process of the system (mass transport, kinetics, or loading) is directly linked to the system response time. The effect is particularly significant during the summer months as the flows are relatively lower compared to other times of the year.

1. Phytoplankton Modeling of Sacramento-San Joaquin Delta

The Sacramento-San Joaquin Delta phytoplankton model was originally developed by Di Toro et al.[36] to study eutrophication in that system. The delta is horizontally divided into 39 segments with no vertical subdivision (see Figure 4 in Chapter 3). In each segment, the following water quality constituents are simulated: phytoplankton chlorophyll *a*, zooplankton carbon, nonliving organic nitrogen, ammonia nitrogen, nitrate nitrogen, nonliving organic phosphorus, orthophosphate, CBOD, dissolved oxygen, chloride, and silica. The kinetic formulations were considered adequate and up-to-date given the data available at the time of study.

To justify seasonal modeling approximations, salinity data are examined in Figure 16. First, the steady state salinity data measured in the physical model under a net delta outflow of 10,000 *cfs* are displyed, showing tidally-averaged, maximum, and minimum concentrations along the main channel of the delta. Second, salinity measurements from two prototype slack surveys in August and September, 1970 are translated into mean tide levels for comparison with the steady-state physical model data. Also shown in Figure 16 is an annual hydrograph for the net delta outflow in 1970. The hydrograph indicates significant spring runoff followed by low summer flows. In 1970, the system would be approaching a steady-state condition by August. The steady-state salinity data from the physical model are bound by the two prototype surveys, another indication that modeling the system during August and September in 1970 would closely approximate the system.[42] Note that the physical model had been calibrated so that its data mimic the field conditions.

To further demonstrate the validity of seasonal steady-state approximation for the eutrophication analysis of the system, seasonal modeling results are compared with the time-variable model results for 1970, 1974, and 1976, three years with different hydrologic and environmental conditions. The comparisons are summarized in Figure 17 for the following key water quality constituents: phytoplankton chlorophyll *a*, nonliving organic nitrogen, nitrate, and dissolved oxygen at Chipps Island.

The monthly average net delta outflows are also presented. The year 1970 is characterized with very high flows (above 150,000 *cfs*) in January followed by a sharp decline during February and March. By April, the flow would drop to 7000 *cfs*. The flows in June and July are even lower and begin a steady rise in August, eventually reaching 85,000 *cfs* by the end of the year. In 1974, the winter flow progressively increases from below 40,000 *cfs* in January to above 107,000 *cfs*

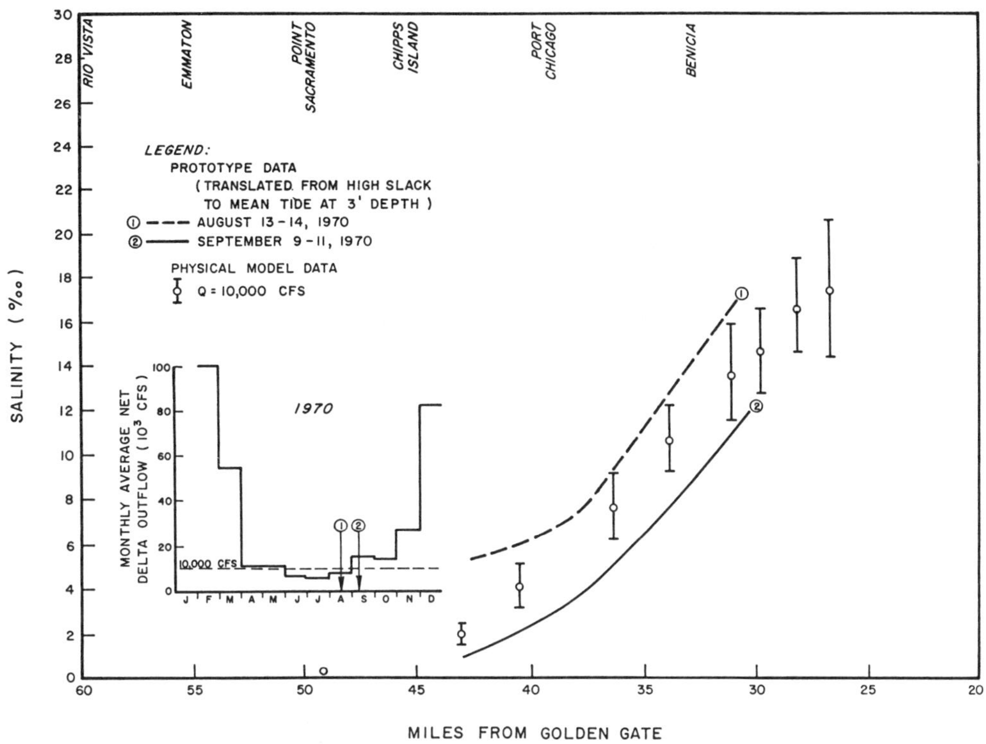

FIGURE 16. Seasonal salinity and freshwater flow variation in Sacramento-San Joaquin Delta.

in April. Again, a sharp decline lowers the flow to 10,000 *cfs* in July, which is followed by a slow rise to above 30,000 *cfs* in December. A drought condition is seen for 1976. The average flow in January of that year is slightly above 9000 *cfs* while the lowest average flow (below 4000 *cfs*) occurs in June. The drought condition persists through the rest of the year. It should be pointed out that temperature is another factor affecting the seasonal divisions.

Figure 17 shows that the results from the seasonal steady-state approximations (winter, spring, summer, and fall) are bound by the time-variable modeling results, closely following the seasonal trends. The spring steady-state calculations yield a chlorophyll *a* peak during the spring months when the phytoplankton biomass reaches its maximum level. The concentrations of nitrogen components are also reproduced by the steady-state analysis, matching the time-variable model results. The nitrate nitrogen level reaches its minimum in spring when the phytoplankton biomass attains its peak. Results from the dissolved oxygen calculations follow the seasonal trend as the dissolved oxygen level is primarily a function of temperature and salinity levels in the system. The comparison between the model results and the observed data is encouraging, considering many approximations in the steady-state analysis.

2. Eutrophication Modeling of the James Estuary

Eutrophication in the Chesapeake Bay and its tributaries is one of the most important environmental, economic, and political problems in recent years. The James River contributes approximately 24 to 36%, (depending on the hydrologic condition in the basin) of the total phosphorus loads to the bay.[43] Prior to 1988, there was no phosphorus control in the James River Basin (see Figure 9).

A modeling study on the upper James Estuary was conducted by Lung[13] to evaluate point source phosphorus control for the James River Basin. The water quality model JMSRV presented in Section IV.B was used to study the factors limiting the phytoplankton growth in the

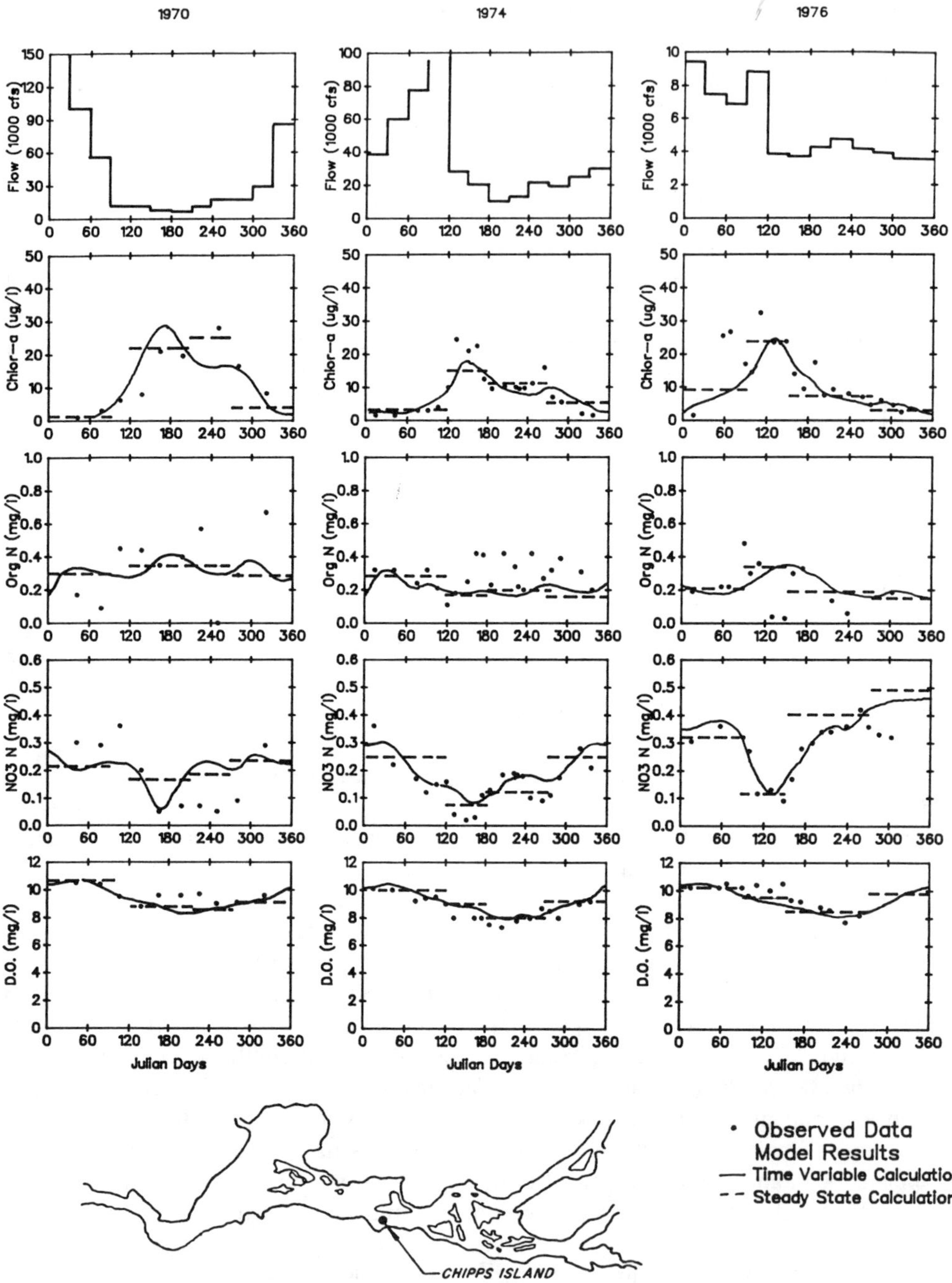

FIGURE 17. Seasonal steady-state eutrophication modeling of Sacramento-San Joaquin Delta.

upper James Estuary. The water quality constituents simulated by the model are CBOD, dissolved oxygen, organic nitrogen, ammonia nitrogen, nitrite plus nitrate nitrogen, organic phosphorus, inorganic phosphorus, and chlorophyll *a*. Kinetics of major physical, chemical, and biological processes that link these water quality constituents are modeled in each of the 50 model segments in the longitudinal direction of the upper James Estuary.

Model calibration results using two separate sets of data sets in the summer of 1983 are summarized in Figure 18. In general, the increase in ammonia concentration below Richmond

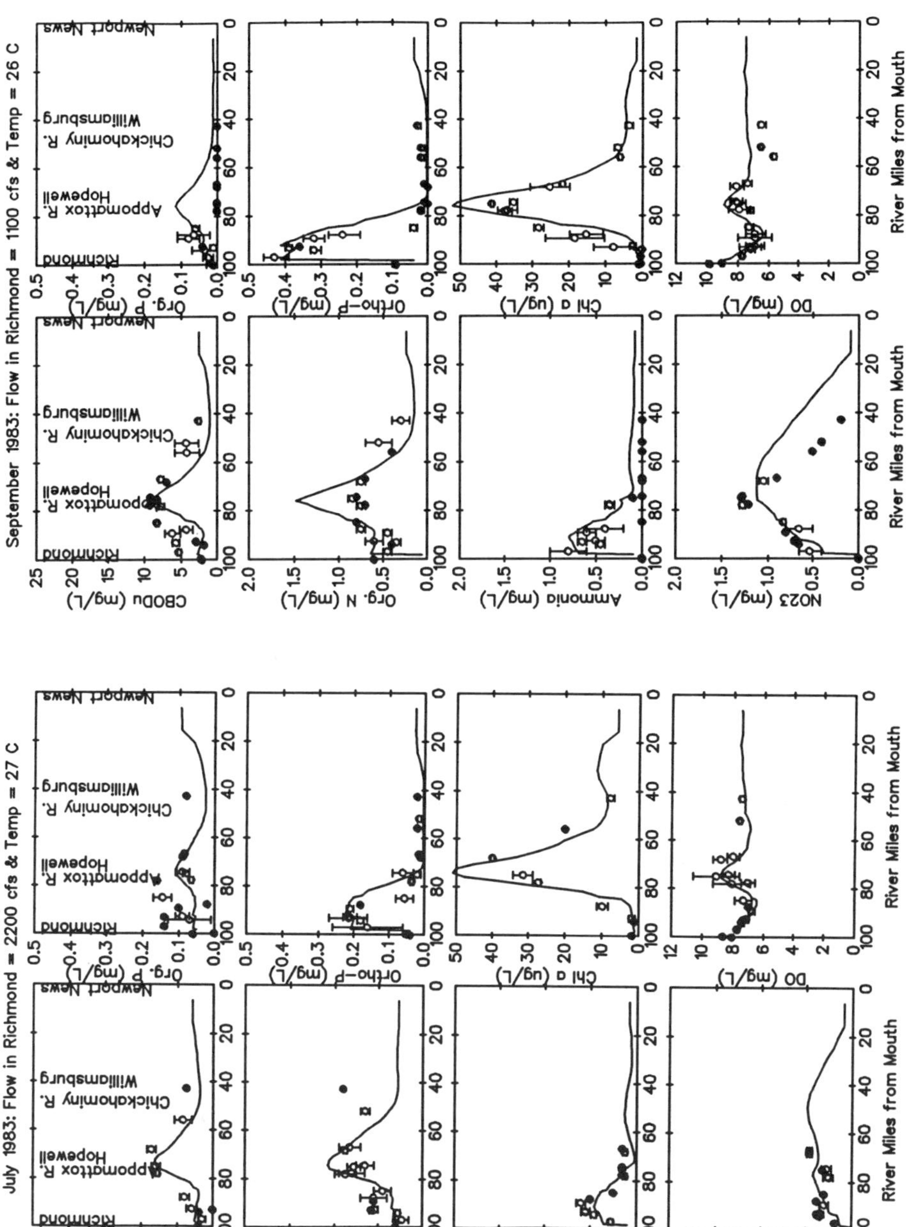

FIGURE 18. Eutrophication modeling analysis of James Estuary (July and September 1983 data).

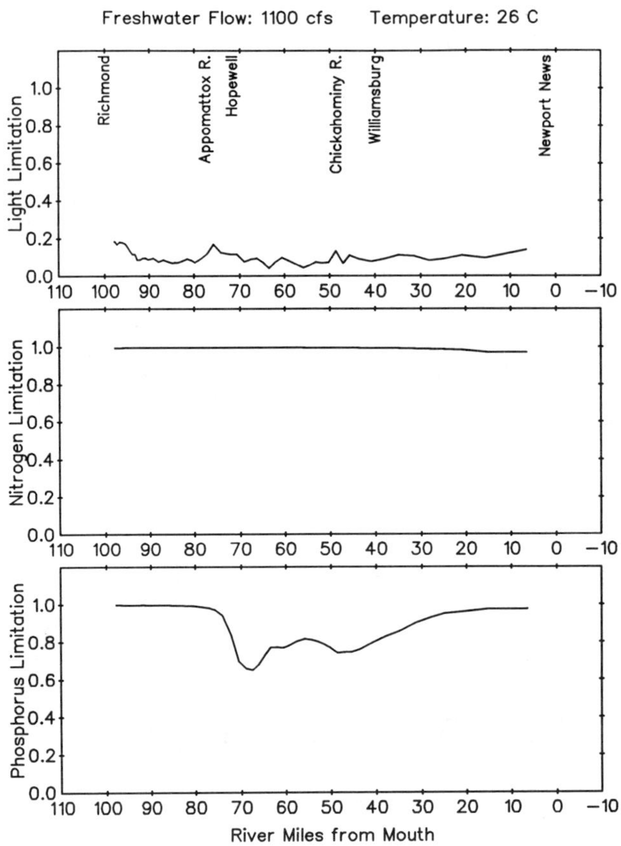

FIGURE 19. Factors affecting phytoplankton growth in the James Estuary.

was due to ammonia discharge from point sources in that area. However, the increase did not sustain beyond Mile Point 90 because of phytoplankton uptake and nitrification. The orthophosphate profile in the upstream area closely resembles the ammonia profile for the similar reason: municipal wastewater input. Subsequent decrease in orthophosphate concentrations was the result of algal uptake and adsorption by suspended particles. The lowest orthophosphate concentration is about 0.01 mg/l, which is still higher than the Michaelis-Menton constant (0.001 mg/l) limiting the algal growth in the model.

Results from the September 1983 model calibration run are further summarized in Figure 19 to provide additional insight into the phytoplankton growth dynamics in the upper estuary. Figure 19 presents light and nutrient limitations on algal growth rate phytoplankton chlorophyll *a* in the James Estuary under a freshwater flow of 1100 *cfs* and at a water temperature of 26°C. Light and nutrient limitations are based on available light intensity and nutrient levels and are expressed as dimensionless numbers in terms of percentage of the optimum algal growth rate. It should be pointed out that average channel depths below Richmond increase for the first 20 mi and then decrease, reaching some shallow sections near the Hopewell area. This local shallowness resulted in favorable light conditions and thereby peak chlorophyll *a* levels near Hopewell. The light extinction coefficient increases progressively in the downstream direction from 1.3 m^{-1} near Richmond to 3.0 m^{-1} by the junction of the Chickahominy River. Strong light attenuation was the primary cause for the algal biomass to decline following the biomass peak. Phytoplankton settling is another reason for the decline of the biomass. Further, light extinction in the upper estuary is high, significantly limiting algal growth in the water column. On the other hand, nutrients (nitrogen and phosphorus) were not limiting factors as their contribution to a less than optimum growth rate was small.[13]

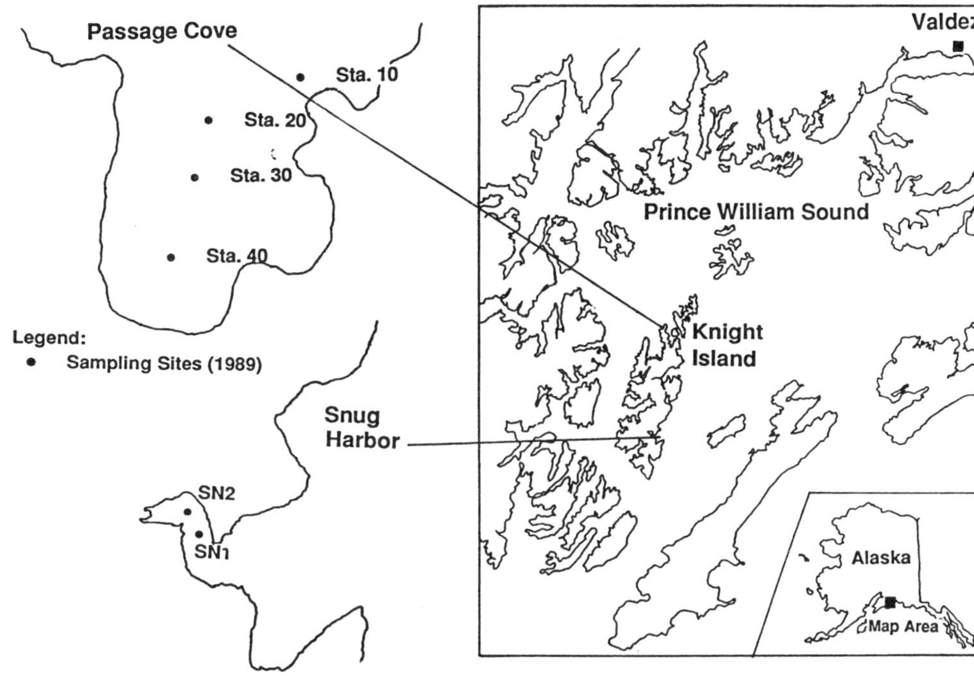

FIGURE 20. Passage Cove and Snug Harbor, Knight Island, Prince William Sound.

Given the above quantitative phytoplankton-nutrient dynamics, the model was used to assess the effect of point source phosphorus control. The control alternatives evaluated include phosphate detergent bans and phosphorus removal at POTWs. Lung[13] showed that phosphate detergent bans would provide small reductions in peak chlorophyll *a* levels in the upper estuary due to the fact that phosphorus concentrations are not low enough to limit algal growth. However, considerable reductions by phosphorus removals would offer more promising benefits in reducing chlorophyll *a* concentrations in the water column.[13]

In 1987, the Virginia General Assembly took action to reduce nutrient enrichment by enacting a phosphate detergent ban. It became effective on January 1, 1988. In March 1988, the Virginia State Water Control Board adopted the Policy for Nutrient Enriched Waters and a water quality standard designating certain waters as nutrient enriched. Under the policy, municipal and industrial wastewater treatment plants permitted to discharge more than 1 mgd to nutrient enriched waters are required to remove phosphorus to meet a 2 mg/l limit. Facilities are given up to three years to complete plant modifications to meet this limit.

3. Eutrophication Potential Modeling of Coastal Embayments

The U.S. EPA bioremediation program to investigate cleaning up the contaminated beaches following the 1989 Exxon Valdez oil spill involved a field demonstration project conducted to determine whether nutrient addition to contaminated beaches would stimulate hydrocarbon breakdown by indigenous bacteria. One of the concerns associated with nutrient application was whether the added nutrients would cause excessive algal growth during the growing season in the embayments where the study was conducted. Because of limited data in the study area, a modeling approach was chosen to address these questions by simulating the eutrophication potential. Two embayments in Knight Island, Passage Cove, and Snug Harbor (Figure 20) were modeled. Snug Harbor is located on the southeastern side of Knight Island. Major sources of freshwater runoff are from precipitation and snowmelt, which is typical of islands in Prince William Sound. Although some shorelines in Snug Harbor were heavily contaminated with oil,

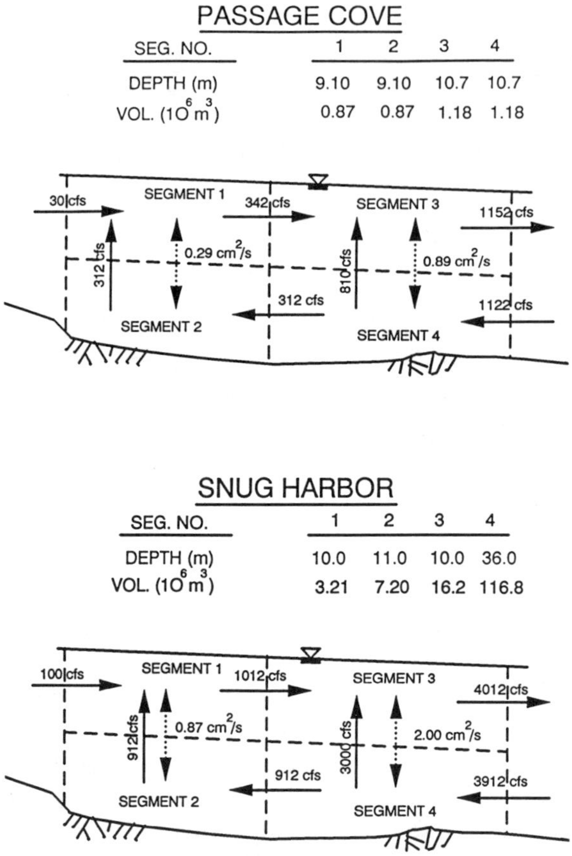

FIGURE 21. Two-layer mass transport in Passage Cove and Snug Harbor.

it appeared that little oil was being released to the water, thus minimizing the prospect of reoiling on the beaches chosen for treatment and reference plots. A second site chosen for treatment was Passage Cove, which is located just northwest of the northern tip of Knight Island (Figure 20). Even after physical washing with hot water, considerable amounts of oil remained at this site, mostly spread uniformly over the surface of rocks and in the beach material below the rocks.

Data and information available for this study are limited from the modeling standpoint. While a brief summary of the available data is presented in the following paragraphs, a more comprehensive description of the field data may be found in another document.[44]

Freshwater runoff from small streams into the embayments could only be estimated. On a seasonal average basis, the freshwater flows from the watersheds were estimated to be about 30 (0.9 m^3/sec) and 100 *cfs* (2.8 m^3/sec) in Passage Cove and Snug Harbor, respectively. These estimates were based on one or two observations of the approximate surface velocities and on crude estimates of cross-sectional areas of the major streams. NOAA navigation charts for Knight Island were examined to calculate the volume and surface area of the two embayments. Average depths were also determined. In addition, the relationship between volume and depth was developed. Tidal fluctuations reach a maximum of 15 ft (4.57 m).

A number of water quality surveys were conducted in Passage Cove and Snug Harbor during the bioremediation study in 1989. Temperature and salinity profiles indicate that the water column is partially mixed in both embayments. Higher temperatures and lower salinity levels in the surface layers (up to 10 m deep) are due to the freshwater inflows. pH levels show little vertical variation. Dissolved oxygen profiles show progressive increase from the surface to the thermocline probably due to decreased temperature. Below the thermocline, bacterial and algal

TABLE 7
Model Calculation (Surface Layer) vs. Field Data

Parameter	Passage Cove	Snug Harbor
Chlorophyll *a* (μg/l):		
Model	0.84–1.13	1.09–1.12
Measured[a]	0.70–0.90	0.90–1.10
Productivity rate (mg C/m³/h):		
Model	1.07–1.50	1.36–1.38
Measured[a]	1.00–2.50	0.20–1.40

[a] Pritchard et al.[45]

respiration and biomass decomposition reduce the dissolved oxygen concentration in the water column.

Chlorophyll *a* concentrations measured in the summer of 1989 in the surface waters of Passage Cove and Snug Harbor range from less than 1 μg/l to slightly over 2 μg/l[45] with the maximum values being observed immediately following the fertilizer application. Algal productivity rates observed during the same period in these two systems are below 1.0 mg C/m³/h prior to the treatment and up to 3.0 mg C/m³/h after the treatment.[45] These are relatively low productivity rates, another indication of insignificant eutrophication. Nitrate concentrations are about 0.012 to 0.025 mg/l; orthophosphate concentrations range from 0.009 to 0.012 mg/l.

The methodology for quantifying two-layer estuarine mass transport, presented in Chapter 3, was used to calculate the two-layer mass transport in the two embayments. Figure 21 shows the mass transport coefficients in a four-segment configuration for Passage Cove and Snug Harbor. Note that the freshwater flow rates are small compared with the two-layer flows. The average salinity calculated for each segment was compared with the measured salinity. Subsequent model sensitivity analysis through iterations of this procedure fine tuned the mass transport coefficients to match the measured salinity data.[44] Based on the mass transport coefficients (advective and dispersive flows) in this four-segment model, the flushing times in these two systems were calculated to be 5.6 and 28.4 tidal cycles for Passage Cove and Snug Harbor, respectively. The flushing times are slightly longer than those determined by the tidal prism method but substantially shorter than those determined solely by freshwater dilution. In summary, the flushing times for both embayments are relatively short, resulting from the tidal actions.

The EUTRO4 model[46] was adapted to the two embayments with site-specific data (such as the mass transport coefficients developed and described in the preceding section) to simulate the seasonal steady-state water quality in the systems. Most kinetic coefficients for algal nutrient dynamics and nutrient recycling were derived from values reported for estuaries and embayments with similar characteristics.[44] A review of the literature did not yield any applicable studies in Prince William Sound. Environmental conditions such as light intensity, water temperature, photoperiod, and light extinction coefficient were estimated for the study area. Boundary conditions for algal biomass and nutrients were developed from data collected from ambient waters in Prince William Sound.

The calculated seasonal algal chlorophyll *a* concentrations and algal production rates in the surface layer of Passage Cove and Snug Harbor are shown in Table 7 along with measured values. Table 7 indicates that model results match the measured algal chlorophyll *a* concentrations and algal productivity rates very well.

A number of fertilizer types and application procedures were tested for application to Prince William Sound. A typical loading estimate for fertilizers sprayed on the test beaches was about 7.0 and 0.7 kg/d for nitrogen and phosphorus, respectively.[44] For the modeling analysis, this

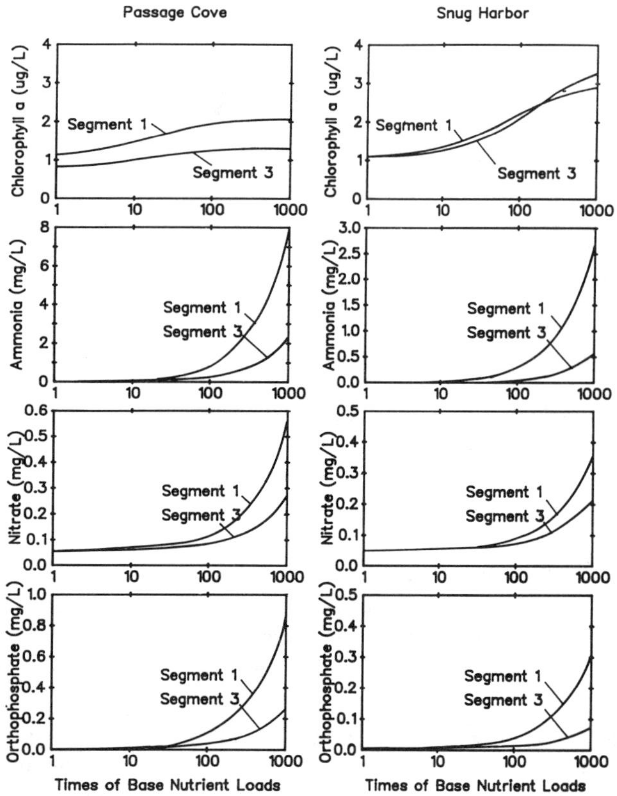

FIGURE 22. Predicted response to nutrient additions in the surface layer of Passage Cove and Snug Harbor.

application rate was multiplied by factors ranging from 0 to 1000 in order to evaluate all possible loading conditions. This loading was assumed to be applied directly to the water column and neglects losses that may have occurred due to uptake by organisms in the beach or nearshore zone.

Model results of nutrient addition are summarized in Figure 22. Total algal chlorophyll *a*, ammonia nitrogen, nitrate nitrogen, and ortho-phosphate concentrations in the surface layer of Passage Cove and Snug Harbor are shown with increasing nutrient input (in terms of factors of base loading rates). Essentially, the model results indicate that the maximum chlorophyll *a* concentrations could reach only 1 to 2 μg/l in Passage Cove and about 2.9 to 3.3 μg/l in Snug Harbor.

To determine how long it would take for the embayments to recover after the nutrient treatment stops, the eutrophication models were run with constant nutrient loading rates for about 100 d when the algal biomass in the embayment approached equilibrium with the nutrient loads. The nutrient loads were then discontinued. The model results showed that depending on the loading conditions (10, 100, and 1000 times the base rate), Snug Harbor would take approximately 20 to 60 d to reduce the nutrient concentrations to the pretreatment levels. In comparison, the recovery time for Passage Cove is much shorter (no more than 15 d regardless of the loading factor) due to its small size.

Significant concentrations of ammonia in the water column were predicted by the model under high application rates of fertilizer (Figure 22). At a water temperature of 12.4°C and pH of 7.5 as used in the model calculation, about 0.75% of the ammonia is un-ionized. As such, the inner Passage Cove would violate the 0.02 mg/l un-ionized ammonia criterion at a loading factor

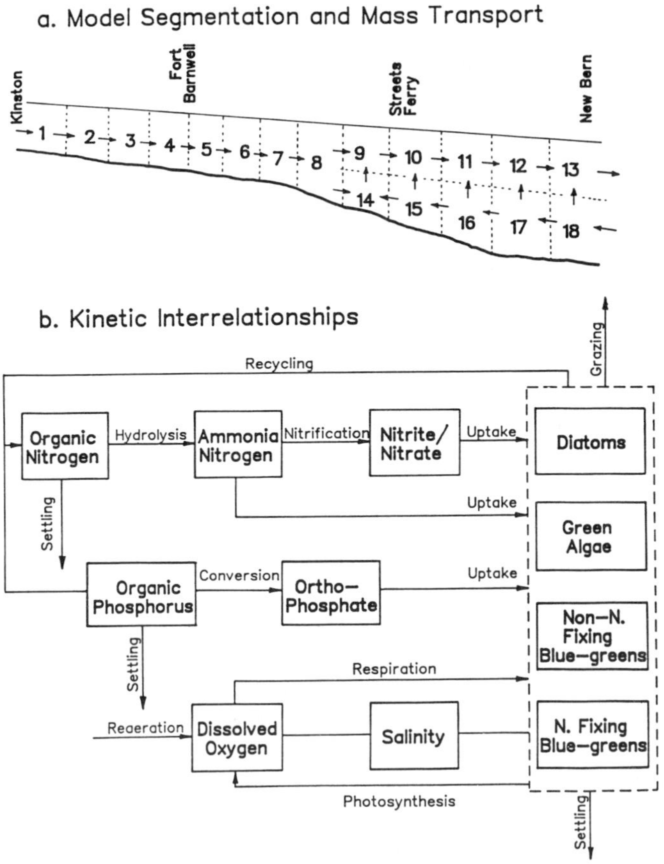

FIGURE 23. Neuse River Estuary model segmentation and water column kinetics.

of about 300. On the other hand, Snug Harbor is not expected to exceed this criterion with the maximum nutrient load rate (1000 times of the base rate).

This case study demonstrated the use of a comprehensive water quality modeling framework to assess the eutrophication potential in two coastal embayments. Although the data was limited, a number of checkpoints were developed along with model sensitivity analyses in the study to guide the model projections. The model results were intended to quantify the incremental impact of nutrient applications on eutrophication. They should not be viewed as actual predictions for algal biomass levels in the embayments.

B. TIME-VARIABLE PHYTOPLANKTON MODELING

While steady-state phytoplankton modeling as described in Section VI.A could provide much insight into seasonal algal growth-nutrient dynamics, it has limitations. In estuarine systems, different nutrients become the limiting factor in different seasons. Further, peak algal biomass levels in the surface waters usually do not coincide with lowest dissolved oxygen in the bottom waters. More importantly, the algal biomass produced in the spring months could contribute to the sediment oxygen demand or nutrient release for the sediment during the summer months. It is necessary to evaluate the time-variable trends of the water quality with respect to estuarine eutrophication. One of the recent time-variable eutrophication models is the Potomac Estuary Model (PEM).[19] PEM has been useful in guiding the water quality management for the Potomac and played an important role in analyzing the 1983 blue-green algal blooms in the estuary.[47] Since the development of PEM, several estuarine eutrophication models have been developed for time-variable computations. The following paragraphs present two recently developed models for nutrient controls.

1. Modeling Blue-Green Algae in the Neuse River

Over the decade, segments of the lower Neuse River between Kinston and New Burn, NC (Figure 23) revealed alarming symptoms of advanced eutrophication, culminating in the appearance and persistence of nuisance blue-green algal blooms.[48,49] Spring-summer-fall blooms of nuisance blue-green genera (*Microcystis*, *Oscillatoria*, *Anabaena* and *Aphanizomenon*) that at times can coat the river with green paint-like scums. On average, such blooms can be expected at a frequency of once every two or three years.

Questions arising during consideration of management options included:

- What are the factors and their interrelationships that cause blue-green algal blooms in the lower Neuse River?
- Would major reductions of nutrient inputs (either from point or nonpoint sources or both) to the lower Neuse River help to arrest the occurrence and persistence of nuisance blue-green algal blooms?

A mathematical model of the lower Neuse River was developed to help evaluate control alternatives. The modeling effort focused on a better understanding of the mechanisms initiating and sustaining algal blooms in the lower Neuse River. Figure 23 shows the model segmentation, mass transport, water quality constituents modeled, and associated kinetics. Multiple functional groups of phytoplankton (diatoms, green algae, nonnitrogen fixing and nitrogen-fixing blue-green algae) are incorporated into the model. In addition, nutrient components and dissolved oxygen are also included in the model. Two-layer mass transport in tidal and estuarine portions of the lower Neuse River is considered to characterize the surface-dwelling blue-greens in the surface layer. Time-variable simulations (tidally averaged) of seasonal phytoplankton and nutrient dynamics, incorporating seasonal variations of mass transport are conducted. A complete description of the model design can be found in Lung and Paerl.[50]

The model (water column kinetics) was calibrated and verified using the data collected in 1983, 1984, and 1985. Model results using the 1983 data are summarized in Figure 24. Both model results and field data indicate that ortho-phosphate is always in ample supply for algal growth throughout the year. Nitrogen supply preceding the algal blooms appears sufficient. During the bloom period, ammonia and nitrate levels are reduced significantly. The two-layer transport patter reproduces the temporal and spatial salinity distributions very well. Figure 25 presents a comparison between calculated and measured chlorophyll *a* levels in 1983 for the four phytoplankton groups. Diatoms are dominant during early spring but are progressively replaced by the blue-greens during the blooms.

The key mechanisms responsible for the seasonal succession of algal groups are due to the different growth dynamics formulated in the model for different algal groups. First, temperature dependence of growth rate is different between the diatoms, greens, and blue-greens (see Figure 13a). At high water temperatures particularly in the late summer months, the blue-greens have a much greater growth rate than the diatoms and greens. Second, a very small settling velocity ranging between 0.05 and 0.19 ft/day is assigned for the blue-greens while the diatoms have a settling velocity of 1.5 ft/day. Although independent measurements of settling velocities were not available, these settling velocity values are within the literature reported range.[2]

To identify the key factor(s) controlling the blue-green algal bloom, Lung and Paerl[50] evaluated the following parameters: freshwater flow, nutrient input, water temperature, and light extinction coefficient (Figure 26). The year 1983 was unique in that the summer flows were extremely low and the summer water temperatures were clearly higher than normal. They also reported that the blue-green algae growth rates were more or less the same in 1983 to 1985. Yet,

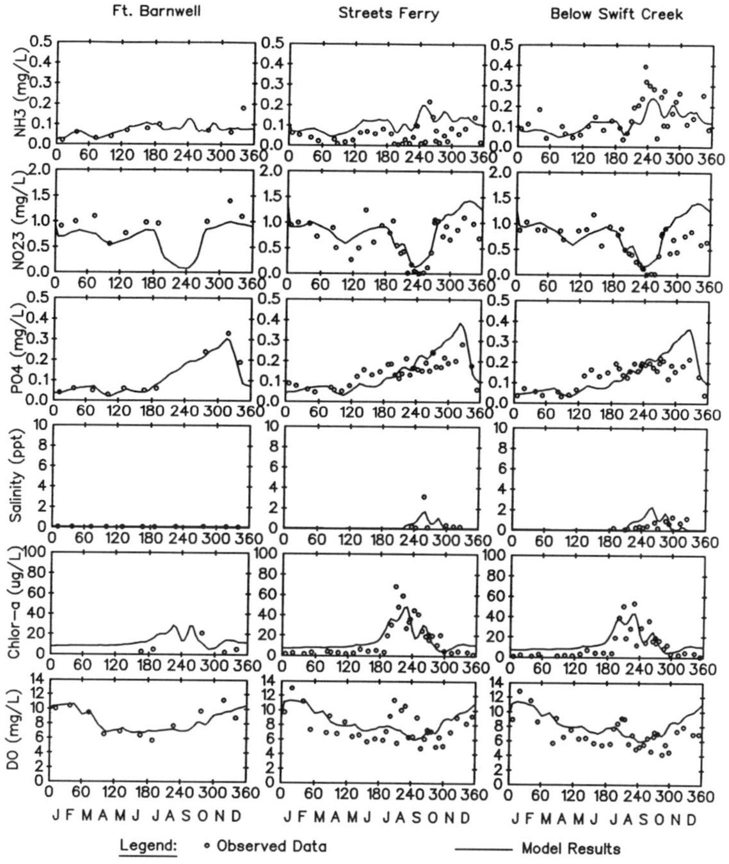

FIGURE 24. Neuse River Estuary eutrophication model analysis.

when the seasonal flow condition effect was incorporated, the model yields different responses between 1983 and 1984, an indication that freshwater flow conditions were directly responsible for the blue-green algae blooms in the lower Neuse River.

Lung and Paerl[50] concluded that initiation of the blue-green algal blooms in the lower Neuse River is strongly regulated by summer river flow under excessive (for growth) nutrient concentrations. Huffman[51] used the model to evaluate a number of nutrient control alternatives for the Neuse River. Point source controls involved a stepwise reduction of effluent concentrations of total phosphorus from 6 to 0.2 mg/l and total nitrogen from 7 to 3 mg/l. Reductions of present nonpoint source nitrogen and phosphorus loads by 30% were also evaluated in conjunction with point nutrient controls. In addition, low-flow augmentation was also examined as a method to limit bloom formation. The model results indicate that a 38% reduction in maximum annual chlorophyll *a* concentrations may be achieved with stringent point source phosphorus control. Nonpoint source nitrogen and phosphorus controls of 30% were effective in limiting growth only when implemented in conjunction with point source phosphorus control. Flow control simulations suggest that periodic release of water from upstream reservoirs may arrest algal growth during low flows.

2. Eutrophication and Dissolved Oxygen Modeling of the Patuxent Estuary

The Patuxent River (see Figure 12 in Chapter 3) has been a major focal point for water quality management in Maryland for well over a decade. At the present time, broad efforts to clean up the Patuxent are substantially underway with measurable progress in implementation of the policies set forth in a 1983 Water Quality Management Plan. Based on a modeling study in the

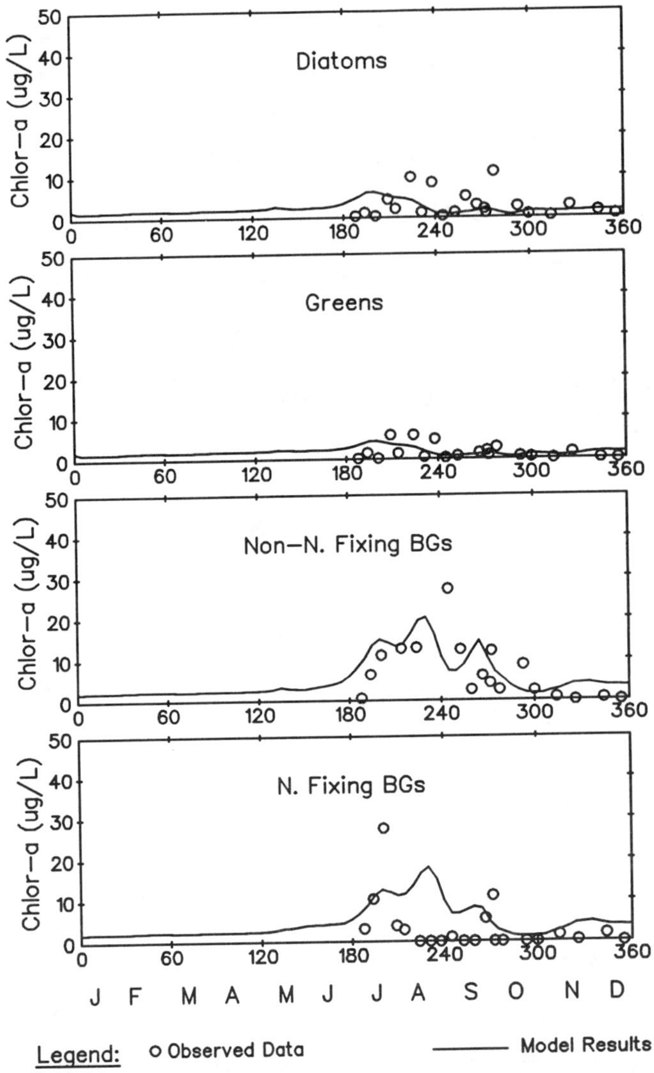

FIGURE 25. Neuse River Estuary eutrophication model results — multiple algal groups.

1981, phosphorus removal has been installed at major wastewater treatment plants with flows greater than 1 mgd in the Patuxent River basin. Additional removal of nutrients (i.e., nitrogen) has been considered for the Patuxent River basin. However, a series of water quality management questions still remain unresolved. These questions are

- What is the quantitative relationship between nutrient loading and eutrophication and anoxia in the lower estuary? More specifically, sediment oxygen demand vs. downstream boundary conditions (i.e., the Bay effect), nutrient fluxes, and degree of vertical stratification should be addressed.
- What are the critical nutrient(s) controlling eutrophication and anoxia, in consideration of the factors involved in algal/nutrient dynamics in the estuary. Further, under what conditions could nitrogen limitation overtake phosphorus limitation? This information is necessary to justify nitrogen removal at the wastewater treatment plants, since phosphorus removal has generally been implemented prior to nitrogen removal.

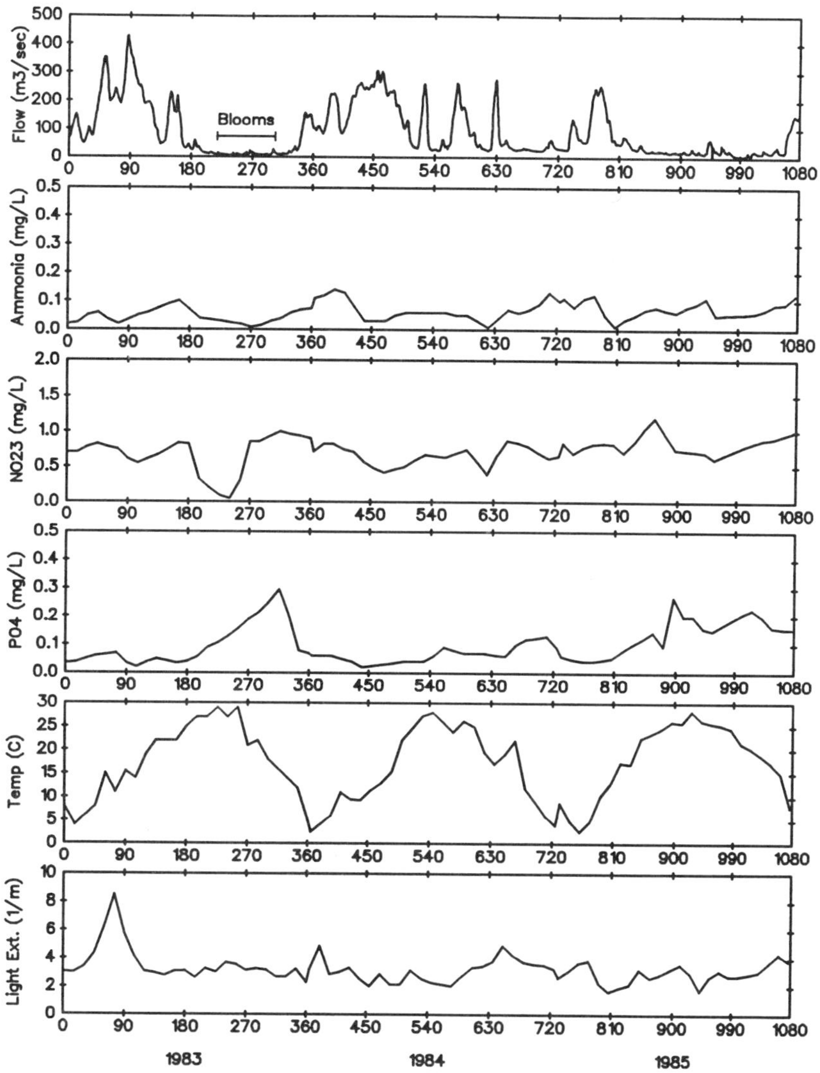

FIGURE 26. Factors affecting blue-green algal blooms in Neuse River Estuary.

- What degree of nutrient control is needed? The trade-off between point and nonpoint source nutrient controls with respect to the eutrophication and anoxia in the Patuxent Estuary must be evaluated.
- How long will it take for the water quality of the Patuxent Estuary to improve once controls are fully implemented? The rate of recovery may be affected by an active sediment layer which could continue to supply nutrients to the system even after point and nonpoint source nutrient controls are achieved. This is an important question which needs to be addressed through long-term model simulations.

To address these questions, the Maryland Department of the Environment (MDE) launched a Patuxent River Monitoring, Research, and Modeling Strategy which includes the validation of a watershed model, an estuarine hydrodynamic model, and an estuarine water quality model. As part of the effort, a comprehensive water quality model was developed.[41] It is a tidally averaged time-variable model which simulates seasonal dynamics of phytoplankton, nutrient uptake/recycle, and sediment-water interactions. Spatially, the water column is sliced into two layers, each of which is subdivided into 19 segments in the longitudinal direction (see Figure 4 in Chapter 3). Additional features of the model include two algal groups (diatoms and green algae) and a sediment layer. The sediment plays a major role in eutrophication of the Patuxent. Spring and summer algal growth in the upper estuary results in subsequent deposition of algal biomass to the sediment where decomposition consumes dissolved oxygen. Depressed oxygen levels have been observed in the bottom water of the lower estuary in the summer. Nutrient (ortho-phosphate and ammonia) releases from the sediment under anaerobic condition represent a significant nutrient source in the estuary. Such sediment-water interactions are incorporated into the model by including a sediment layer (with 19 segments) in addition to the water-column layers.

Water quality constituents modeled include chlorophyll *a* for the two algal groups, zooplankton carbon, nitrogen components (organic nitrogen, ammonia, and nitrite/nitrate), phosphorus components (nonliving dissolved and particulate organic phosphorus, dissolved and particulate inorganic phosphorus), dissolved oxygen, and total suspended solids in the water column and sediments. Sediment nutrient fluxes are calculated in terms of concentration gradients of the nutrients between the water column and interstitial water. Sediment oxygen demand is calculated in a similar manner. The water column kinetics processes included in the model are quite similar to those in Figure 12.

The water quality model was then run continuously from 1983 to 1985. The mass transport patterns for these three years were assembled to drive the water quality model (see Chapter 3). Model kinetic coefficient values were kept the same throughout the three-year simulation period. Only external variables such as solar radiation, photoperiod, wind speed, water temperatures, and boundary conditions were modified accordingly for 1983, 1984, and 1985.

Results of the 3-year continuous run are summarized in temporal plots for two key locations: Stations PXT0402 (Nottingham) where phytoplankton growth is persistent, and XDE5339 (Broomes Island) where the bottom water becomes anoxic during the summer months. Figure 27 shows the comparison between model results and field data for the following water quality parameters at Nottingham. The water quality parameters presented are total nitrogen, ammonia, nitrite/nitrate, total phosphorus, ortho-phosphate, chlorophyll *a*, and dissolved oxygen. In Figure 27, the data indicates that total nitrogen concentrations in the upper estuary is scattered and are closely influenced by the nitrogen input at the fall line. The model results show less temporal variations throughout this three-year period. The model results match the measured total phosphorus and ortho-phosphate concentrations very well. Model-calculated chlorophyll *a* concentrations also match the data, although the model underestimates the peaks in 1985. Dissolved oxygen concentrations at these locations are strongly influenced by phytoplankton growth, in addition to the seasonal variation due to temperature changes in the water column.

The same water quality variables at Broomes Island (Station XDE5339) are presented in Figure 28. First, total nitrogen, ammonia, and nitrite/nitrate concentrations are reproduced by the model, following their seasonal variations closely. Next, measured total phosphorus and ortho-phosphate concentrations are reproduced very well for the three-year period. Sediment releases of orthophosphate in the summer months are shown. Chlorophyll *a* levels at these locations are considerably lower than those observed at Nottingham and are accurately calculated by the model. Finally, the dissolved oxygen concentrations show that the vertical stratification of dissolved oxygen and the temporal extent of the hypoxic conditions are predicted by the model.

Longitudinal profiles of water quality constituent concentrations calculated by the model are compared with the data. Again, the same water quality constituents for spring 1983 are summarized in Figure 29. The data plotted for comparison in each plot are in layer-averaged values from water quality surveys over any given period. In general, the model reproduces the data very well. Three key features of the Patuxent Estuary water quality are accurately mimicked by the model: high nutrient concentrations near the head of the estuary, persistent phytoplankton

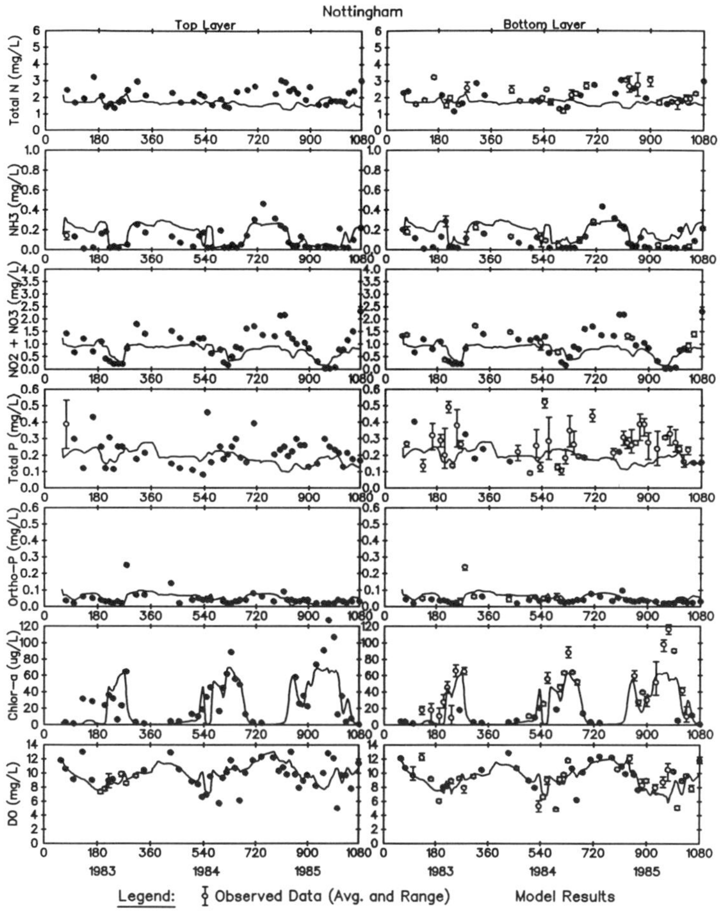

FIGURE 27. Patuxent Estuary water quality model results — Nottingham (1983–1985).

biomass at Nottingham, and low dissolved oxygen levels in the bottom waters at Broomes Island.

3. Assigning Kinetic Coefficients

Formulating the water column kinetics and assigning the kinetic coefficients and parameters require much empiricism for many reasons. In many cases, data for independent derivation of kinetic coefficients is either limited or not available, making the task almost impossible. In practice, while independent estimates of the exogenous variables such as freshwater flow rates, preliminary mass transport patterns, boundary conditions, and environmental conditions could be readily derived from the data, model kinetic coefficients derived from literature values have been validated via the model calibration process. It should be stressed that model calculations are not intended to curve fit the data. Rather, the calibrated coefficients (e.g., algal growth kinetics) are fine tuned through a series of model sensitivity runs with reasonable and usually narrow ranges of their values developed from the literature and other estuarine modeling studies. Water quality data collected in other years with different environmental conditions would further enhance the validity of the model such as in the Neuse and Patuxent Estuary modeling studies.

It should also be pointed out that in model sensitivity analyses, adjusting the kinetic coefficients and constants (even within their narrow ranges) to improve the calibration of certain water quality constituent(s) often would result in adverse outcome of matching other water

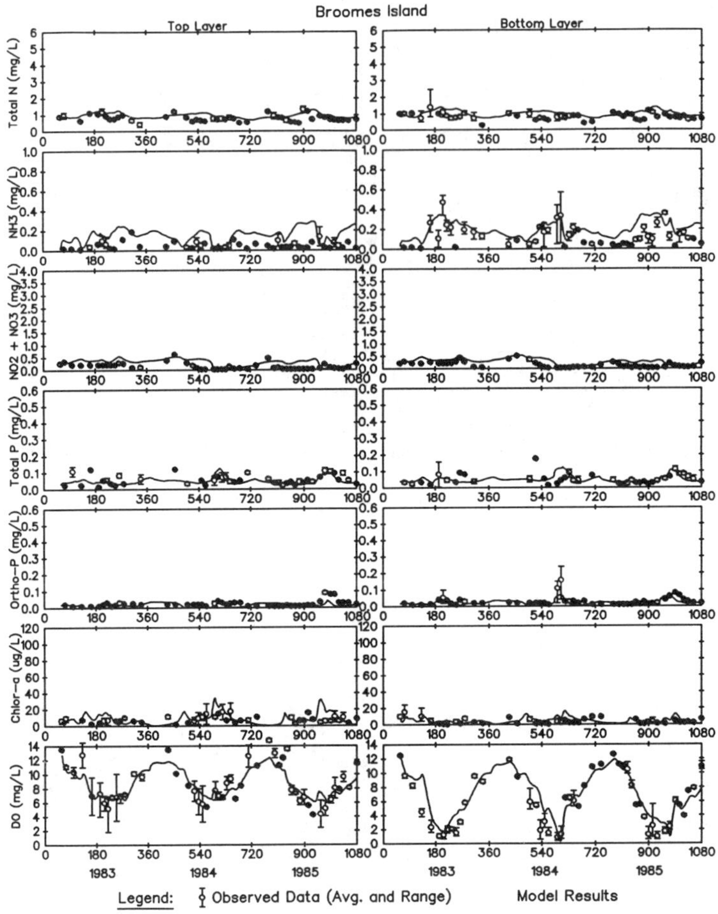

FIGURE 28. Patuxent Estuary water quality model results — Broomes Island (1983–1985).

quality constituents or kinetic processes. These are the constraints that a modeler would face in the model calibration process. Such exercises would eventually lead to the development of a unique set of model coefficients. The key kinetic coefficients in recent estuarine eutrophication models are compiled and listed in Table 8 for comparison.

VII. NUTRIENT CONTROL AND WATER QUALITY MANAGEMENT

Nutrient reductions have been practiced in water quality management for many estuarine systems. The achievement can be grouped into short-term and long-term water quality benefits. The first sign of the water quality improvement is the reduction of the nutrient concentrations in the estuarine system, which in turn would reduce the algal biomass. A successful nutrient control program should provide another benefit: a progressive reduction of nutrient releases from the sediment and the associated sediment oxygen demand. The following sections discuss the technical aspects in estuarine water quality modeling that are closely linked to the decision-making for water quality management.

A. LIMITING NUTRIENT

The important question that arises in water quality management is should inputs of phosphorus or nitrogen or both nutrients be controlled? One simple approach to this question is

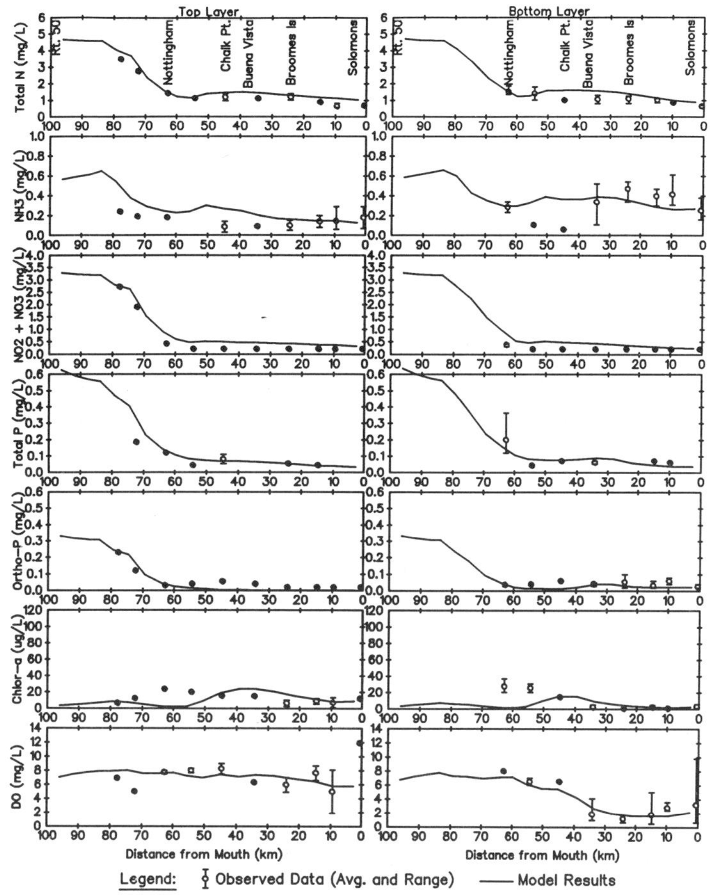

FIGURE 29. Patuxent Estuary water quality model results — longitudinal profiles.

to examine the relative nutrient requirements by plants for nitrogen and phosphorus (the *N*/*p* ratio) using Figure 14 which plots the nitrogen concentration against the phosphorus concentration in the water column, with a ratio of 10 as the threshold for nutrient control. Thomann and Mueller[1] presented a guideline to assess the controlling nutrients in estuaries. One should be reminded that this simple concept of using the *N*/*p* ratios is subject to some variation due, for example, to the variability in algal stoichiometry.[1] As a result, *N*/*p* ratios of 20 or greater generally reflect phosphorus limited systems while *N*/*p* ratios of 5 or less reflect nitrogen limited systems.

Figure 30 shows the ratios of dissolved inorganic nitrogen (ammonia and nitrite/nitrate) to dissolved ortho-phosphate concentrations in the layer of the water column in the Patuxent Estuary from 1983 to 1985. The data indicates a potential of phosphorus limitation in the spring followed by nitrogen limitation in the summer as suggested by D'Elia et al.[52] The summer ratios in the middle estuary at stations XED4892 and XDE9401 are consistently below ten, suggesting nitrogen limitation.

The above simple concept is particularly useful in assessing the potential of controlling nutrient(s). A key question in water quality management is how much nutrients can be discharged to the body of water so that some desired level of algal biomass is maintained? The models described in this chapter are useful tools to provide answers to these questions. Many eutrophication models to date use the Michaelis-Menton formulation to quantify the significance of nutrient limitation. Further, in multiple nutrient cases, such as nitrogen and phosphorus,

TABLE 8
Water Column Kinetic Coefficients and Constants in Estuarine Eutrophication Models

Parameter	Unit	Chesapeake[a]	Potomac[b]	Patuxent[c]
Algal Sat. Growth Rate (20°C)	d^{-1}	2.0–2.5	2.0	2.0 (Diatoms) 1.8 (Green Algae)
Temperature Coefficient, θ	—	1.068	1.068	Diatoms: Gm = 0.1* T for 0°C ≤ T≤30°C Gm = –0.6* T + 21.0 for 30°C ≤ T ≤ 35°C Green Algae: Gm = 0.0 for 0°C ≤ T ≤ 5°C Gm = 0.12* T – 0.60 for 5°C ≤ T ≤ 35°C Gm = –0.36* T + 16.2 for 35°C ≤ T ≤ 40°C
Optimum Light Intensity	ly/d	350	300	350 (Diatoms and Green Algae)
Algal Resp Rate (20°C)	d^{-1}	0.1–0.125	0.125	0.125
Temperature Coefficient, θ	—	1.047	1.045	1.080
Algal Death Rate (20°C)	d^{-1}	0.1–0.2	0.02	0.125
Temperature Coefficient, θ	—	1.047	1.045	1.080
Algal Settling Velocity	ft/d	0.33	0.30	1.0 (Diatoms) 0.3 (Green Algae) 1.0 (Non-living Part. Org. P) 1.0 (Part. Inorg. P) 0.5 (Total Org. N) 0.5 (CBOD) 0.5 (Inorg. Suspended Solids)
Phosphorus Partition Coef.	l/kg			12800
Michaelis Constant (P)	μg/l	1.5	1.0	1.0
Michaelis Constant (N)	μg/l	15.0	25.0	25.0 (Diatoms and Green Algae)
Carbon/Chlorophyll *a*	mg C/mg Chl *a*	30–50	40	50
Nitrogen/Chlorophyll *a*	mg N/mg Chl *a*	8–17.5	10	7
Phosphorus/Chlorophyll *a*	mg P/mg Chl *a*	0.55–1.23	1	1
Oxygen/Chlorophyll *a*	mg O/mg Chl *a*	80–133	107	133.5
Org N Hydrolysis Rate (20°C)	d^{-1}	0.03	0.075	0.02
Temperature Coefficient, θ	—	1.08	1.08	1.08
Nitrification Rate (20°C)	d^{-1}			0.03
Org P Conversion Rate (20°C)	d^{-1}	0.03	0.22	0.02

Temperature Coefficient, θ	—	1.08	1.08	1.08
Denitrification Rate (20°C)	d^{-1}			0.09
Temperature Coefficient, θ	—	1.08	1.08	1.08
Half-Saturation Rate	mg DO/l			0.50
Recycled Fraction of P in Algal Biomass	—	0.1	0.1	0.1 (to DOP)
		0.4	0.4	0.4 (to POP)
		0.5	0.5	0.5 (to DIP)
Recycled Fraction of N in Zooplankton Biomass	—	0.75	0.75	0.75 (to TON)
BOD_u/BOD_5		3.0	1.85	3.0
CBOD Decay Rate (20°C)	d^{-1}	0.08	0.21	0.2
Zooplankton Grazing Rate	l/d/mg C			1.8
Half Saturation Rate	mg Chl *a*/l			0.05
Yield Coefficient	mg C/mg C	0.70	0.70	0.70
Optimum Grazing Rate	d^{-1}	1.0	1.0	1.0
Zooplankton Resp. Rate	d^{-1}	0.05	0.05	0.05
Temperature Coefficient, θ	—	1.08	1.08	1.08

[a] HydroQual, Inc., Development of a Coupled Hydrodynamic/Water Quality Model of the Eutrophication and Anoxia Processes of the Chesapeake Bay, 1987.
[b] Thomann, R.V. and Fitzpatrick, J.J.[19]
[c] Lung.[41]

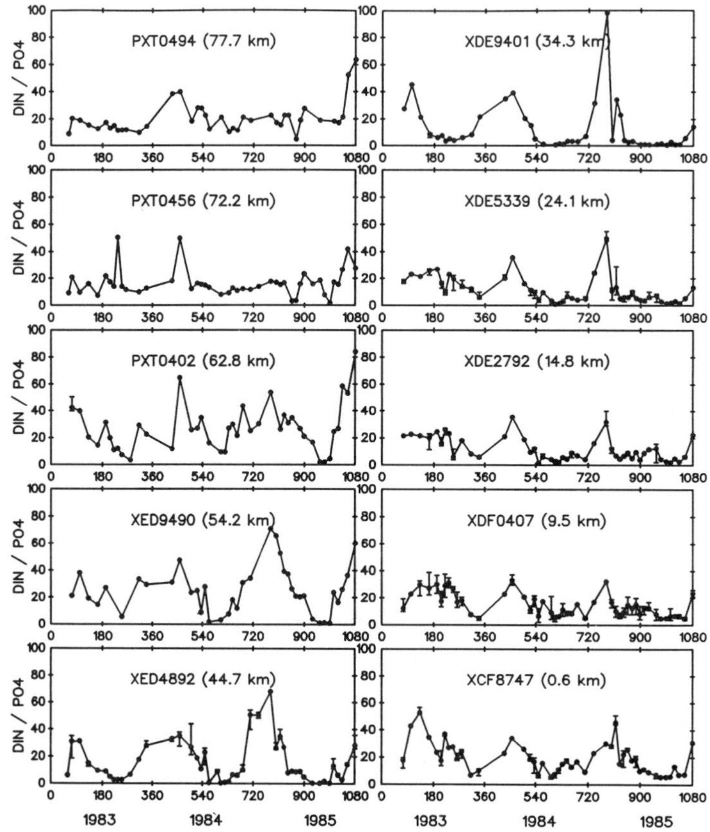

FIGURE 30. DIN/DIP ratios in top layer of the Patuxent Estuary (1983–1985).

the minimum formulation is used. That is, the factor for the shortest supply will control the growth of algae. Compounding the nonlinear relationship in the Michaelis-Menton formulation and the minimum formulation yields one of the most important features of eutrophication kinetics in water quality management.

To justify nitrogen removal for the Patuxent Estuary which already has a phosphorus control (point sources) strategy in place, the Patuxent Estuary model developed by Lung[41] was used to quantify the reduction of the biomass in the upper estuary where significant algal bloom occur consistently. Figure 31 shows time-variable plots of nitrogen limitation factors in the top layer at different locations along the Patuxent Estuary. To overcome the phosphorus limitation factor under existing conditions, both point and nonpoint nitrogen input needs to be reduced significantly with the assumption that ammonia releases from the sediment in the lower estuary would be curtailed due to external nitrogen load reductions.

It should be pointed out that there are other factors that could affect the nutrient limitation concept. For example, luxury uptake of phosphorus by certain algal species would significantly change the nutrient limitation factor based on the Michaelis-Menton equation. Thus, caution must be exercised in interpreting the data on this issue.

B. SINGLE VS. DUAL NUTRIENT CONTROL — A CRITICAL MANAGEMENT DECISION

In the Potomac and Patuxent Estuaries, phosphorus control was first implemented. Over the past two decades, a staged reduction in effluent phosphorus concentrations for the wastewater treatment plants around the watersheds has been observed, resulting in significant improvement

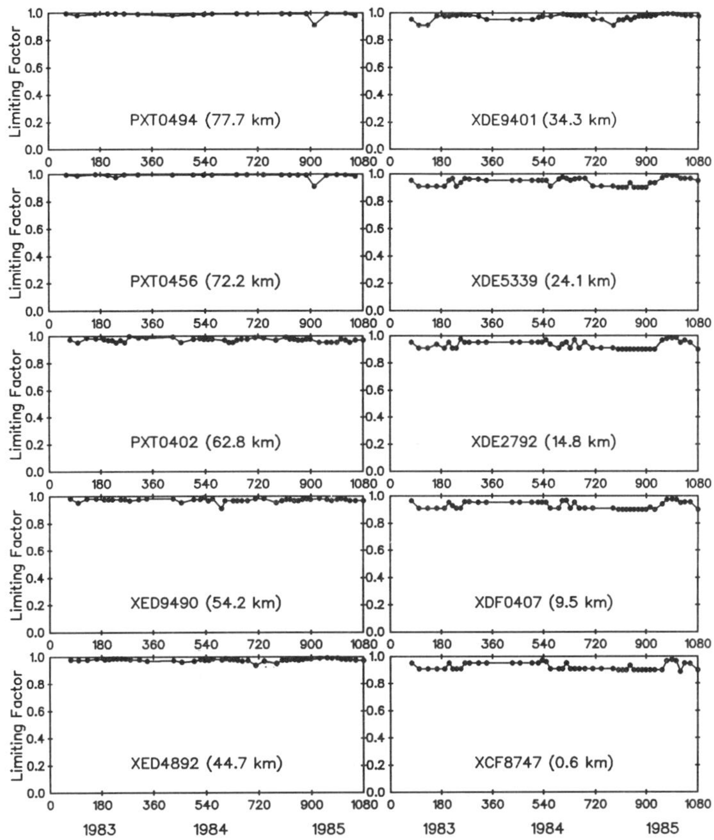

FIGURE 31. Nitrogen and phosphorus limitation in top layer of the Patuxent Estuary (1983–1985).

of the water quality. It has been argued that in freshwater habitats phosphorus input constraints will ultimately be the only effective step in stemming eutrophication. In theory this is true, since forcing any system into a phosphorus-limited condition should allow water quality managers to arrest and regulate algal productivity. In more practical terms, phosphorus removal is much easier to achieve than nitrogen removal on the economical basis. In reality, however, sediments may be an important, internal source of phosphorus to continue supporting the algal growth. One of the excellent examples is the Potomac Estuary.[47] After phosphorus removal was well underway, a major phytoplankton bloom occurred in 1983 causing considerable concern. The expert panel that investigated the bloom concluded that the prevailing weather and flow conditions through the end of July 1983, in combination with available nutrients in the water column led to the development of the bloom. The bloom was intensified and maintained during the period of August through October due to phosphorus releases from river bottom sediments. Interestingly, releases of phosphorus from sediments were not attributed to anoxic conditions. Instead, pH levels of 9 and above in overlying waters were suspected as the principal driving force that stimulated increased phosphorus release from bottom sediments and intensified the bloom beyond its initial stage.[47] Nevertheless, the Potomac case study clearly pointed out the need to quantify the sediment-water nutrient interactions in a eutrophication model.

C. NITROGEN CONTROL AND NITROGEN FIXING PHYTOPLANKTON

The Neuse River Estuary has been observed with repeated blue-green algal (*Microcystis aeruginosa*) blooms. Clearly, it can be seen that variable hydrological conditions, particularly evident as wet vs. dry spring-summer months, play a crucial role in determining bloom potentials in respective years.[50,53] Both adequate nutrient (specifically nitrogen) loading during wet spring months as well as subsequent low discharge (low flushing) conditions in summer/fall months play joint roles in dictating the magnitudes and persistence of blooms. Optimal bloom development occurred in 1983, a year that featured both excessive winter-spring nutrient loading as well as dry, low discharge summer months. Other years, such as 1982 and 1984, featured adequate (excessive) spring nutrient loding; however, continued high discharge in summer-fall months negated bloom potentials. Conversely, 1985 and 1986, while exhibiting record droughts and resultant extensive low summer discharge conditions, failed to support detectable nuisance blooms. These years also yielded extremely dry, low discharge spring periods, thus minimizing associated nutrient loading during the months of February/May. Consequently, inadequate (for subsequent *Microcystis aeruginosa* bloom development) nutrient loading took place in these years.

Clearly, a reduction in nitrogen loading, to a level below *Microcystis's* bloom threshold would lead to reduction, if not total elimination, of this organism's bloom potential. However, nitrogen fixing analog surface bloom genera, including *Anabaena* and *Aphanizomenon*, could potentially dominate under nitrogen-limited conditions intended to eliminate *Microcystis* blooms. Both nitrogen fixing genera are currently present as subdominants in the Neuse River summer phytoplankton community. The Potomac Estuary, which from physical, hydrological, and alkalinity perspectives are similar to the Neuse River, periodically reveals shifts within the blue-green algal community, away from *Microcystis* domination to dominance by nitrogen fixing genera as inorganic nitrogen depletion is established during mid- to late summer months.[53]

It would therefore seem inescapable that parallel control of potential nitrogen fixing blue-green algal blooms must be considered for the Neuse River if we are to successfully eliminate all potential nuisance genera. Such control measures would necessitate phosphorus input reductions. The rationale for this projection is simple. Blue-green algae capable of nitrogen fixation have the biological options of either shutting down their nitrogen fixing apparatus in the event of inorganic nitrogen availability or excess (current Neuse River conditions), or utilizing nitrogen their nitrogen fixing apparatus when inorganic nitrogen is deficient but the remaining essential nutrients are sufficient. Thus, the possibility exists that inorganic nitrogen removal alone would simply cause a shift within the blue-green algal community, away from non-nitrogen fixers toward nitrogen fixers of equal notoriety. Therefore, identification and reduction of phosphorus inputs to the extent that they would maintain potential nitrogen fixers below a bloom threshold is crucial in formulating an effective nutrient management strategy for the Neuse River basin.[49]

D. LONG-TERM WATER QUALITY RESPONSE

One of the important management questions to be addressed in eutrophication control is how long will it take the estuarine system to improve its water quality once controls are implemented? More specifically, would the sediment continue to supply nutrients to and demand oxygen supplies from the water column? In other words, how dynamic (active) is the sediment? There are no easy answers to these questions. A case in point is the Potomac Estuary in Washington, D.C. Phosphorus removal was well under way in the late 1970s and early 1980s. In fact, point-source phosphorus loadings had been reduced to loadings comparable to the 1930s. Average effluent concentrations were about 0.4 mg$P$/l. Then in the summer of 1983, a major bloom developed of the blue-green alga *Microcystis aeruginosa*. The occurrence of algal blooms in the Potomac in spite of extensive phosphorus reduction indicates the complexity of the eutrophication of estuaries.[47]

The key to the answers lies in the sediment. It is reasonable to expect that under continuing point and nonpoint source nutrient controls, the sediment fluxes of oxygen demanding material and nutrients would respond to these controls and be diminished after a period of time. In early eutrophication modeling studies, modelers used judgement to assign the percent reduction (eventually achieved) in sediment oxygen demand and sediment fluxes of nutrients following external nutrient load reductions. The question of how long would it take to reach that stage was not addressed.

To address these questions, the eutrophication model must consist of a sediment system that interacts with the water column on a continuing basis, thus requiring a time-variable model. Several technical difficulties would develop in such a modeling effort. (More comprehensive discussions of this topic can be found in Chapter 6.) One of the recent modeling studies including this effort is the Chesapeake Bay model in which a two-layer sediment system is incorporated. Basically, the top layer is aerobic and the second layer is anoxic. Diagenesis of organic materials is considered. Another case study is the Patuxent Estuary[41] in which the sediment system is formulated following that in the Chesapeake Bay model.

E. FATE OF NUTRIENTS: NUMERICAL TAGGING

It is known that nutrients once discharged into the receiving waters, such as an estuary, would find their way into a variety of "compartments" in the estuarine system. For example, phosphorus from wastewaters in an estuary could be incorporated into the biomass of phytoplankton in the water column, deposited into the sediments, or transported to the lower estuary. Perhaps a more meaningful question is how much phosphorus in the algal biomass at a certain location in the water column is from an identified source? Component analyses are routinely performed to quantify the contribution of individual sources to dissolved oxygen deficits as many dissolved oxygen models have a linear relationship between BOD loads and DO deficits (see Figure 8). A similar component analysis is not appropriate for eutrophication modeling analysis because of the nonlinear nature of the phytoplankton growth-nutrient dynamics in the model. That is, results from a component analysis would not predict algal biomass accurately in terms of the various sources of phosphorus without taking into consideration other factors that also control algal production. Lung and Testerman[54] pointed out the erroneous results from a simple component analysis in eutrophication modeling. The sum of the algal biomass generating from individual phosphorus sources is greater than the biomass (as chlorophyll *a*) from the model calibration. The same argument is true when predicting the reduction of algal biomass following the reduction of the limiting nutrient in the estuarine system. Thus, a different approach, called *numerical tagging*, is needed to address this issue.

In limnological studies, $^{32}PO_4$ is added as a tracer to determine the fate of phosphorus in systems by measuring the amount of ^{32}P in various components of the system. The concept of using ^{32}P as a tracer can be adopted in a mathematical fashion in eutrophication modeling. That is, a source or sources of phosphorus can be numerically labeled and introduced into the estuary. The eutrophication model can then be used to trace the amount of such labeled phosphorus in different compartments of the estuarine system: particulate and dissolved phosphorus.

Lung and Testerman[54] presented an example of such an analysis for the James Estuary under steady-state conditions. The objective is to determine the amount of phosphorus from individual sources in organic phosphorus, ortho-phosphate, and algal biomass. First, the kinetics of the James Estuary model are modified to include three more system variables: labeled organic phosphorus, labeled ortho-phosphate, and labeled phosphorus in the algal biomass. As such, parallel calculations of labeled and unlabeled phosphorus are incorporated into the model. Special care is needed to treat the nonlinear relationship between algal growth rate and phosphorus concentrations. While the phosphorus limiting level is still calculated based on the sum of labeled and unlabeled ortho-phosphate concentrations, the amount of labeled and unlabeled ortho-phosphate incorporated into the algal biomass is proportional to the ortho-phosphate concentrations available in the water column. When either labeled or unlabeled ortho-phosphate is exhausted, the contribution of that ortho-phosphate to biomass ceases. Such a practice is necessary to maintain the mass balance for phosphorus and also to avoid generating negative ortho-phosphate concentrations by the model. Other kinetic interrelationships between these labeled system variables are the same as those for unlabeled variables.

The modified model has been tested for conservation of mass as well as numerical accuracy. For example, when a single source of phosphorus is labeled, the modified model would still generate the same total organic phosphorus, ortho-phosphate, and algal biomass levels as the original model calculated. Next, five categories of phosphorus input: Richmond POTW, other POTWs, industrial input, upstream and nonpoint input, and Appomattox River to the James Estuary are labeled one at a time. Each time, the total organic phosphorus, ortho-phosphate, and phosphorus in the algal biomass are added and found to equal the total phosphorus concentrations calculated by the original unlabeled model.

Results from this analysis using the calibration data set of September 1983 are presented in Figure 32. The POTWs in the upper James Estuary contributed about 75% of the total algal biomass (as chlorophyll *a*) in the water column. The Richmond wastewater treatment plant is a major contributor of ortho-phosphate as well as algal biomass in the upper James Estuary. Input from upstream (primarily nonpoint) and downstream boundaries provided another 15% of the algal biomass in the upper estuary. Industrial wastewaters played a very small role in contributing to the algal biomass in the James Estuary. The Appomattox River, which receives wastewater discharges from the city of Petersburg, contributed an insignificant amount of phosphorus to the algal biomass in the mainstream of the James Estuary.

VIII. MODELING TOOLS

One of the most commonly used computer programs in water quality at the present time is WASP (Water Quality Analysis Simulation Program). WASP is a generalized modeling framework for modeling contaminant fate and transport in surface waters.[55] It is a versatile program, capable of studying time-variable or steady-state, one-, two-, or three-dimensional, linear or nonlinear kinetic water quality problems. The model is based on the box model approach. To date WASP has been employed in many modeling applications, that have included river, lake, estuarine, and ocean environments and that have investigated dissolved oxygen, bacterial, eutrophication, and toxic substance problem contexts. Some recent WASP applications to estuarine eutrophication modeling studies include the Potomac Estuary by Thomann and Fitzpatrick,[19] the Neuse Estuary by Lung and Paerl,[50] and the Patuxent Estuary by Lung.[41]

The program is flexible enough to provide the modeler with the mechanisms to describe the kinetic processes and the inputs to these processes, as well as the transport processes and the geophysical morphology or settling, that go into the framework of the model. Transport processes, basically hydrodynamic in nature, include advection, turbulent diffusion, and, when spatial averaging is included, dispersion. Kinetic processes are the sources and sinks that act upon a particular water quality parameter and may be physical, chemical, or biological. The user needs to formulate the kinetics in the water column and sediment pertinent to the water body being modeled and thus has the flexibility to include additional water quality constituents in the model. Although the development of the subroutine is a burden to the modeler, it gains flexibility.

In recent years, EPA's Center for Exposure Assessment Modeling (CEAM) in Athens, Georgia, has adopted and modified the program. WASP4 is the latest of a series of modifications at CEAM.[46] EUTRO4 is the component of WASP4 which is applicable to modeling eutrophication. The program simulates the transport and transformation of up to eight water quality constituents associated with eutrophication and dissolved oxygen in the water column and sediment bed. While limiting the maximum number of water quality constituents to eight, EUTRO4 eliminates the need to write a kinetics subroutine. That is, the programming task has been completely taken away from the user who would also have options of five levels of complexities in kinetics to be modeled.

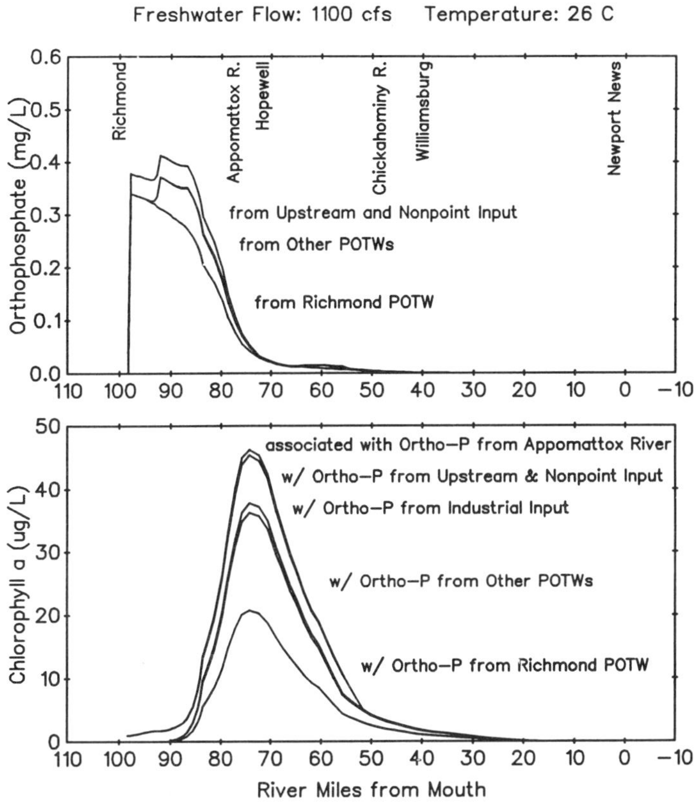

FIGURE 32. Numerical tagging modeling results — James Estuary.

With the rapid advancement of computer hardware and software in recent years, many estuarine water quality modeling computations can be performed on microcomputer systems. In fact, most of the application examples presented in this chapter were executed on a 486 system. Simple computations such as the steady-state analysis for the James Estuary and Metawoman Creek are very fast (see Table 9). The time-variable calculations for the Neuse Estuary and Patuxent Estuary are much more involved. Not only do they use time-variable time steps (ranging from 0.05 to 0.15 d), they also incorporate time-variable mass transport patterns, boundary conditions, and inputs. Thus, the computational effort associated with these two models is more extensive. However, the 32-bit FORTRAN compiler significantly increases the execution speed of these models. For example, a three-year run of the Patuxent Estuary model takes about 35 min. on a 486 66 MHz microcomputer. Yet, the same model run would take 20 min of CPU time on an IBM RS 6000 mainframe computer. Such a comparison further demonstrates that the microcomputers are very competitive in performance when compared with the mainframe computers.

Eutrophication models for multiple water quality constituents generate a considerable amount of numerical output for time-variable runs. Thus, processing and interpreting the model results in an efficient manner is important to the modeling analysis. Equally important is comparing model results with observed data for the water quality constituents. Lung[56] has demonstrated graphic displays to show the model results and observed data on the computer system. Essentially, two steps of display are involved. The first is displaying the water quality variables (model results and data) on the monitor screen. Special accommodations are made in the WASP program to process and condense the results for plotting (either spatial or temporal trend). Hardcopies of the screen graphics can be readily obtained on a dot-matrix or laser printer. While the model calibration is assisted with the rapid screen display, the hardcopies of the screen

TABLE 9
Estuarine Eutrophication Models on a 486 66 MHz Microcomputer[a]

Model application	Number of segments	Number of variables	Simulation period	Computation time[b]
JMSRV,				
James Estuary	108	11	120 d[c]	2 min[e]
Metawoman Creek	8	11	120 d[c]	10 sec
EUTRO4,				
Passage Cove	4	8	100 d[c]	5 sec
Snug Harbor	4	8	100 d[c]	5 sec
WASP,				
Neuse Estuary	18	11	1 yr[d]	3 min
Patuxent Estuary	57	16	1 yr[d]	10 min

[a] Using a 32-bit FORTRAN compiler and Weitek Coprocessor.
[b] Does not include post-processing time.
[c] Time-variable run to steady-state under constant loading and boundary conditions.
[d] Tidally averaged time-variable run.
[e] Including screen graphics to display model results.

lack good resolution. The second step of the display is using a plotter to generate quality plots for final presentation in reports or papers.

Finally, it should be pointed out that the computer models, such as WASP and EUTRO4, are only tools. No matter how efficiently they can be executed on computers, they can never replace water quality modeling skills, such as assigning kinetic coefficient values and interpreting the model results. It is therefore important to fully comprehend the water quality problem and to understand the structure of the kinetics for a successful modeling study.

REFERENCES

1. **Thomann, R.V. and Mueller, J. A.,** *Principles of Surface Water Quality Modeling and Control*, Harper & Row, New York, 1987, chaps. 6 & 7.
2. **Bowie, G. L., Mills, W. B., Porcella, D. B., Campbell, C. L., Pagenkopf, J. R., Rupp, G. L., Johnson, M., Chan, P. W. H., Gherini, S. A., and Chamberlin, C. E.,** Rates, Constants, and Kinetics Formulations in Surface Water Quality Modeling (2nd edition), U.S. Environmental Protection Agency, Athens, GA, EPA/600/3-85/040, 1985, chaps. 3, 5, 6.
3. **Di Toro, D. M.,** Documentation of Chesapeake Bay sediment flux model, report prepared by HydroQual, Inc. for U.S. Army Engineers Waterways Experiment Station, 1991.
4. **U.S. EPA,** Technical Guidance Manual for Performing Waste Load Allocations, Book II: Streams and Rivers, Chapter 1: Biochemical Oxygen Demand, Dissolved Oxygen, and Nutrients, 1991.
5. **Di Toro, D. M.,** Algae and dissolved oxygen, Summer Institute in Water Pollution Notes, Manhattan College, Bronx, NY.
6. **Thomann, R. V.,** Systems analysis in water quality management — a 25 year retrospect, in *Systems Analysis in Water Quality Management*, M.B. Beck, Ed., Pergamon Press, London, 1987, 1.
7. **Eckenfelder, W. W. and O'Connor, D. J.,** *Biological Waste Treatment*, Pergamon Press, London, 1961, 4.
8. **Leo, W. M., Thomann, R. V., and Gallagher, T. W.,** Before and after case studies: comparisons of water quality following municipal treatment plant improvements, report submitted by HydroQual, Inc. to EPA Office of Water Programs Operations, Washington, D.C., 1984.
9. **U.S. EPA,** Handbook of Advanced Treatment Review Issues, Office of Water Program Operations, Washington, D.C., 1984, 2.
10. **Metcalf and Eddy,** *Wastewater Engineering*, 3rd ed., McGraw-Hill, New York, 1991, 3.
11. **Hall, J. C. and Foxen, R. J.,** Nitrification in BOD_5 test increases POTW noncompliance, *J. Water Pollut. Control Fed.*, 55, 1461, 1983.

12. **O'Connor, D. J., Di Toro, D. M., and St. John, J. P.,** Water quality analysis of the Delaware River Estuary, *J. Sanit. Eng., ASCE*, 94, 1225, 1968.
13. **Lung, W. S.,** Assessing phosphorus control in the James River basin, *J. Environ. Eng., ASCE,* 112, 44, 1986.
14. **Limno-Tech, Inc.,** Two-Dimensional DEM Model of the Delaware Estuary, report prepared for the Delaware River Basin Commission, 1984.
15. **O'Connor, D. J.,** Oxygen balance of an estuary, *J. Sanit. Eng. Div., ASCE*, 86, SA3, 35, 1960.
16. **Harleman, D. R. F., Dailey, J. E., Thatcher, M. L., Najarian, T. O., Brocard, D. N., and Ferrara R. A.,** User's Manual for the M.I.T. Transient Water Quality Network Model — Including Nitrogen Cycle Dynamics for Rivers and Estuaries, R.M. Parsons Laboratory for Water Resources and Hydrodynamics, Cambridge, MA, 1977.
17. **Cerco, C. F.,** Estimating estuarine reaeration rates, *J. Environ. Eng., ASCE*, 115, 1066, 1989.
18. **O'Connor, D. J.,** Wind effects on gas-liquid transfer coefficients, *J. Environ. Eng., ASCE*, 109, 731, 1983.
19. **Thomann, R. V. and Fitzpatrick, J. J.,** Calibration and Verification of the Potomac Estuary Model, HydroQual, Inc., final report prepared for the Washington, D.C. Department of Environmental Services, 1982.
20. **Hartman, B. and Hammond, D. E.,** Gas exchange in San Francisco Bay, *Hydrobiologia*, 129, 59, 1985.
21. **Thomann, R. V.,** Mathematical model for dissolved oxygen, *J. Sanit. Eng. Div., ASCE*, 89, SA5, 1, 1963.
22. **Pence, G. D., Jeglic, J. M., and Thomann, R.V.,** Time-varying dissolved oxygen model, *J. Sanit. Eng. Div., ASCE*, 94, SA2, 381, 1968.
23. **Jeglic, J. M.,** DECS III, Mathematical Simulation of the Estuarine Behavior, Analysis Memo No. 1032, General Electric Co., Philadelphia, PA, 1966.
24. **Clark, L. J., Ambrose, R. B., and Crain, R. C.,** A Water Quality Modelling Study of the Delaware Estuary, technical report No. 62, EPA Region III, Annapolis Field Office, EPA 903/9-78-001, 1978.
25. **Feigner, K. and Harris, H. S.,** Dynamic Estuary Model Documentation Report, Federal Water Quality Control Administration, Washington, D.C., 1970.
26. **Lung, W. S.,** The Delaware Estuary: a case study of the water quality quality benefits of secondary waste treatment, manuscript prepared for Tetra Tech, Inc., 1991.
27. **Clark, L. J., Ambrose, R. B., and Tortoriello, R. C.,** Documentation Report of the Two-Dimensional Model Water Quality Study of the Delaware Estuary, prepared by the Delaware River Basin Commission, Trenton, NJ, 1982.
28. **HydroQual, Inc.,** Water Quality Analysis of the James and Appomattox Rivers, report prepared for Richmond Regional Planning District Commission, 1966.
29. **Lung, W. S.,** The James Estuary: a case study of the water quality benefits of secondary waste treatment, manuscript prepared for Tetra Tech, Inc., 1991.
30. **Auer, M. T. and Canale, R. P.,** Phosphorus uptake dynamics as related to mathematical modeling at a site on Lake Huron, *J. Great Lakes Res.*, 6,1,1, 1980.
31. **Canale, R. P. and Vogel, A. H.,** The effects of temperature on phytoplankton growth, *J. Environ. Eng. Div., ASCE*, 100,EE1,231, 1974.
32. **Eppley, R. W.,** Temperature and phytoplankton growth in the sea, *Fish. Bull.*, 70, 4, 1063, 1972.
33. **Di Toro, D. M.,** Algae and DO, Summer Institute in Water Pollution Control, Manhattan College, Bronx, NY, 1981.
34. **Di Toro, D.M.,** Optics of turbid estuarine waters: approximations and applications, *Water Res.*, 12, 1059, 1978.
35. **U.S. EPA,** Technical Guidance Manual for Performing Waste Load Allocations, Book III: Estuaries, Part 1: Estuaries and Waste Load Allocation Models, 1990.
36. **Di Toro, D. M., O'Connor, D. J., and Thomann, R. V.,** A dynamic model of the phytoplankton population in the Sacramento-San Joaquin Delta, in *Advanced Chemistry Series*, American Chemical Society, Washington, D.C., 106, 131, 1971.
37. **Bierman, V. J. and Doland, D. M.,** Modeling of phytoplankton in Saginaw Bay: I. calibration phase, *J. Environ. Eng., ASCE*, 112, 400, 1986.
38. **Najarian, T. O., Kaneta, P. J., Taft, J. L., and Thatcher, M. L.,** Manasquan Estuary Study, report prepared for the Manasquan River Regional Sewage Authority, 1981.
39. **Najarian, T. O., Kaneta, P. L., Taft, J. L., and Thatcher, M. L.,** Application of nitrogen-cycle model to Manasquan Estuary, *J. Environ. Eng., ASCE,* 110, 190, 1984.
40. **Thomann, R. V.,** Overview of Potomac Estuary Modeling Tasks I and II — Dissolved Oxygen and Eutrophication, report submitted by HydroQual, Inc. to the Potomac Studies Technical Committee, Washington, D.C., 1980.
41. **Lung, W. S.,** Development of a water quality model for the Patuxent Estuary, in *Coastal and Estuarine Studies, Vol. 36: Estuarine Water Quality Management*, Michaelis, W., Ed., Springer-Verlag, Berlin, 49, 1990.
42. **Lung, W. S. and O'Connor, D. J.,** Assessment of the Effects of Proposed Submerged Sill on the Water Quality of Western Delta-Suisan Bay, report submitted by Hydroscience, Inc. to U.S. Army Engineers Sacramento District, California, 1978.
43. **Lung, W. S.,** Phosphorus loads to the Chesapeake Bay: a perspective, *J. Water Pollut. Control Fed.*, 58, 749, 1986.

44. **Lung, W. S., Martin, J. L., and McCutcheon, S. C.,** Eutrophication and mixing analysis of embayments in Prince William Sound, Alaska, *J. Environ. Eng.*, manuscript accepted for publication, 1993.
45. **Pritchard, P. H., Glasser, J., Kremer, F., Rogers, J., Venosa, A., Chianelli, R., Hinton, S., Prince, R., McMillen, S., and Requejo, A.,** *Oil Spill Bioremediation Project Interim Final Report*, U.S. EPA, Gulf Breeze, FL, 1989.
46. **Ambrose, R. B., Wool, T. A., Connolly, J. P., and Schanz, R. W.,** WASP4, a Hydrodynamic and Water Quality Model — Model Theory, User's Manual, and Programmer's Guide, EPA/600/3-87/039, 1988.
47. **Thomann, R. V., Jaworski, N. J., Nixon, S. W., Paerl, H. W., and Taft, J.,** The 1983 Algal Bloom in the Potomac Estuary, report submitted by the Algal Bloom Expert Panel to the Potomac Strategy State/EPA Management Committee, Washington, D.C., 1985.
48. **Tedder, S. W., Sauber, J., Ausley, J., and Mitchell, S.,** Working Paper: Neuse River Investigations 1979, report prepared by the North Carolina Department of Natural Resources and Community Development, Raleigh, NC, 1980.
49. **Paerl, H. W.,** Dynamics of Blue-Green Algal (*Microcystis aeruginosa*) Blooms in the Lower Neuse River, North Carolina: Causative Factors and Potential Controls, University of North Carolina Water Resources Research Institute Report UNC-WRRI-87-229, 1987.
50. **Lung, W. S. and Paerl, H. W.,** Modeling blue-green algal blooms in the lower Neuse River, *Water Res.*, 22, 895, 1988.
51. **Huffman, L. G.,** Evaluating Eutrophication Control Alternatives for the Lower Neuse River, North Carolina, M.S. thesis, University of Virginia, 1988.
52. **D'Elia, C. F., Sanders, J. G., and Boynton, W. R.,** Nutrient enrichment studies in a coastal plain estuary: phytoplankton growth in large-scale, continuous cultures, *Can. J. Fish. Aquat. Sci.*, 43, 397, 1986.
53. **Paerl, H. W.,** Environmental Factors Promoting and Regulating N_2 Fixing Blue-Green Algae Blooms in the Chowan River, NC, report No. 176, University of North Carolina Water Resources Research Institute, 1982.
54. **Lung, W. S. and Testerman, N.,** Modeling fate and transport of nutrients in the James Estuary, *J. Environ. Eng., ASCE*, 115, 978, 1989.
55. **Di Toro, D. M., Fitzpatrick, J. J., and Thomann, R. V.,** Water Quality Analysis Simulation Program (WASP) and Model Verification Program (MVP) Documentation, report submitted by Hydroscience, Inc. to EPA Environmental Research Laboratory, Duluth, MN, 1982.
56. **Lung, W. S.,** Water quality modeling using personal computers, *J. Water Pollut. Control Fed.*, 59, 909, 1987.

Chapter 5

WATER COLUMN KINETICS II: TOXIC SUBSTANCES

I. INTRODUCTION

This section deals with the kinetics of toxic substances. Like BOD/DO and eutrophication kinetics, comprehensive coverage of mathematical formulations and coefficients of toxic substance kinetics for a variety of water quality problems is readily available from the literature.[1-3] Only a brief discussion of the kinetics is presented in this chapter. As in Chapter 4, case studies of model applications are presented.

II. REVIEW OF KEY PROCESSES

A quantitative understanding of chemical reaction rates and interphase exchanges is critical in predicting the fate and transport of contaminants within a biogeochemical system. The term kinetics as used here refers to the mathematical description of the time-dependency of any dynamic process, both physicochemical (e.g., adsorption, volatilization) and biochemical (e.g., growth, uptake, and biotransformations). The kinetic structure of any model component or system may be designed using either a steady-state, equilibrium, or thermodynamic model or a nonequilibrium (kinetic) model. Each toxicant of concern requires its own kinetic parameters (e.g., rate constants, exchange, or partitioning coefficients), ideally determined under field conditions or estimated from experiments designed with conditions similar to the environment being modeled.

The steady-state condition is time-invariable and implies that the constituent attains equilibrium quickly relative to other kinetic or transport processes which affect that constituent. Steady-state models are relatively easy to design in that they require fewer parameters to describe the process. Experimental data quantifying various equilibrium conditions (e.g., adsorption, speciation) for many conventional pollutants, particularly heavy metals, are available, though only for simple chemical systems. Chemical thermodynamic models are developed on a sound theoretical basis using an extensive database for compound entropies (S), enthalpies (H), Gibbs free energies (G), and equilibrium constants (K_{eq}).

Predicting a nonequilibrium condition is complicated by reaction path dependency, the lack of state functions, such as H or G, and the scarcity of rate constant and activation energy data needed for kinetic models.[4,5] Lack of information on operative reaction mechanisms in the modeled environment often precludes reaction kinetics modeling.

The chemical and physical processes and the level of complexity to be considered in modeling each kinetics structure depend on the level of quantitative understanding of the process and availability of uniformly collected field measurements of sufficient spatial and temporal distribution for model calibration and verification. Generally, the latter restriction is much more severe and limiting in large scale modeling efforts.

Kinetics-controlled processes affecting toxics fate and transport may be grouped into two broad categories: interphase exchange and chemical reactions. The fate and transport of any one compound may be influenced significantly by one or more of the processes listed below. Interphase exchange has no appreciable effect on the chemical structure of the contaminant and includes processes, such as

- Adsorption/desorption to mineral, colloidal, and organic particulates (chemisorption, ion exchange, physisorption)

- Bioaccumulation and bioconcentration
- Air-water exchange (e.g., volatilization, diffusion, sea-salt aerosols)

Chemical reactions involve fundamental change in the molecular structure of the compound and usually its environmental behavior. The reactions may be reversible or irreversible and include the following processes:

- Chemical speciation: the formation of aqueous complexes with inorganic or organic ligands, acid-base and oxidation/reduction (redox) reactions
- Biotransformation (typically microbially mediated)
- Photolysis
- Hydrolysis
- Abiotic reactions such as dissolution/precipitation
- Sediment diagenesis

While these processes and mechanisms are important to the fate and transport of contaminants in the estuarine environment, most contaminants are associated, to a greater or lesser degree, with suspended and deposited particles that are present in estuarine systems. A brief review of these key processes is presented in the following paragraphs.

A. ADSORPTION TO SUSPENDED AND BED SOLIDS

Adsorption of metal and organic toxicants to particulate substrates is one of the most important processes affecting their fate, transport, and bioavailability. Adsorption is dependent on complex electrochemical interactions between solute(s), substrate(s), and solvent(s). The intensity of solute-substrate interaction is ultimately controlled by the nature of the attractive coulombic forces and solute activity. The magnitude of solute-substrate interactions range from negligible as in the case of highly hydrophobic organics (see section below), to relatively weak physisorption, to the relatively strong interaction at specific sorption sites involving chemisorption or ion exchange.

Mechanistic sorption models provide a theoretical basis for interpreting observed sorption behavior. Such models can become quite complicated, depending on the characterization of the water-solid interactions (e.g., electric double layer; see Stumm and Morgan[4]) and sorption site diversity (rate constants for each substrate present, or some fewer, semi-empirical, cumulative rates). Application of highly detailed sorption models are not justified for direct application in estuary-scale, environmental modeling in which substrates are heterogeneous and a wide range of environmental conditions (ionic strength, pH, Eh, competing ion effects, etc.) are anticipated.

Particulate substrates of greatest interest in toxics transport modeling include clay minerals and inorganic (metal oxides and hydroxides) and organic colloids, the colloids occurring as discrete grains or as grain coatings.[6] The particulates may be single grains or aggregates in the form of flocculant or fecal pellets. These substrates are important because they typically have slightly negative surface charges and thus act as cation exchange bases. Additionally, these materials tend to form small particles with a high surface-volume ratio, thus they exert a disproportionate influence on adsorption relative to common coarser grained sediment minerals such as quartz and feldspars, which also have lower exchange capacity.

Both equilibrium and kinetic adsorption models have been developed and applied successfully to simple systems (constant pH and ionic strength, single or few sorbate and sorbent species). Equilibrium models invoke the local equilibrium assumption (LEA), implying relatively rapid adsorption and attainment of a steady-state partitioning. If sorption kinetics is not rapid relative to other kinetic or transport phenomena or the numerical simulation time step, predicted concentrations may be erroneous. For example, if the LEA is imposed on a nonequilibrium system in which the rate of adsorption exceeds desorption, dissolved concentrations are underestimated and sorbed toxicant concentrations are overestimated.

The following discussion of adsorption will address first equilibrium models described by adsorption isotherms. Kinetic adsorption models are discussed briefly. The problematic and important case of adsorption of highly hydrophobic organic adsorption is treated separately.

1. Equilibrium Models: Adsorption Isotherms

Adsorption isotherms describe the distribution of species between the aqueous phase and the water-solid interface over a range of sorbate concentrations at a constant temperature. The isotherms are also strongly dependent upon the nature of the sorbate(s) and substrates(s), and environmental parameters such as pH, Eh, and ionic strength. Observed adsorption behavior may be consistent with a number of isotherms, the most commonly employed being the simple linear, the Langmuir, and Freundlich isotherms. These isotherms are largely empirical.

The simple linear isotherm describes the equilibrium concentration of sorbed constituent (R; typically in μg contaminant per kilogram sediment) as directly proportional to the equilibrium, aqueous concentration (C; typically in μg/l) over the range of concentrations considered:

$$R = K_p C \tag{5.1}$$

where K_p is the partitioning coefficient (or distribution coefficient, K_d, in l/kg). The conceptual and mathematical simplicity of the strictly empirical K_p is appealing, though it has little basis in theory. Extrapolation of the isotherm to concentrations beyond the linear range can yield overestimates of R, thus the isotherm is strictly applicable only at low concentrations of ionic or polar species. The solvophobic behavior of hydrophobic organics tend to follow a linear sorption isotherm, as discussed below.

The Langmuir isotherm describes the adsorption process for the reversible formation of a monolayer of noninteracting sorbate species on a homogeneous population of sorption sites. The nonlinear form of the Langmuir isotherm may be represented by:

$$R = \frac{CR_c}{\frac{K_2}{K_1} + C} \tag{5.2}$$

where

C = aqueous concentration at equilibrium
R_c = maximum adsorption capacity
K_2, K_1 = desorption and adsorption rate constants, respectively

At low concentrations the relation between aqueous and adsorbed concentrations is linear. As the adsorption sites become filled, the isotherm becomes nonlinear with R approaching R_c asymptotically.

The Freundlich isotherm allows for heterogeneity in surface site energies of interaction, described by a Boltzmann type distribution, and may be represented by:

$$R = K_f C^{\frac{1}{n}} \tag{5.3}$$

where

K_f = Freundlich partition coefficient
n = an empirically derived constant

Both the Langmuir and Freundlich isotherms are empirically determined and have some theoretical basis for the stated conditions. Empirical results consistent with these ideal isotherms should not be interpreted as confirmation of the assumptions involved in the isotherm derivation.

2. Kinetic Adsorption Models

The LEA should not be invoked *a priori*, though such practice is common, particularly in when kinetics data are incomplete or unavailable. Nonequilibrium conditions may persist due to chemical and/or physical phenomena in an open system such as an estuary. Chemical nonequilibrium is typically maintained by irreversible phenomena such as decay, volatilization, or dissolution/precipitation reactions. Physical nonequilibrium is usually related to diffusion-controlled reactions which are typically completely reversible.

Wu and Gschwend[7] developed a radial diffusive penetration model to describe the diffusion-limited adsorption kinetics that may be operative in aggregate particulates. Adsorption to suspended aggregates (flocculants, fecal pellets, etc.), which are common in soils and estuarine suspended sediments, may require a nonequilibrium structure.

Karickhoff[8] proposed a mathematical, two-site adsorption model to describe sorption of three polynuclear aromatic hydrocarbons (PAHs) — pyrene, phenanthrene, and naphthalene — on suspended sediments. The model is pseudo-first-order with the "fast" sorption sites which fill relatively quickly (seconds to minutes) and "slow" (hours to days) sites which retard the approach to equilibrium.

Selim and Amacher[9] have developed the second-order, two-site (SOTS), kinetic adsorption model ADKIN, which shows good agreement with observed Cr(VI) transport through porous media. The model characterizes a soil medium as having two sorption site populations, the proportions and rate constants of which are user-specified. Sorption sites in this model are divided into populations which approach equilibrium quickly (seconds to hours; type I) and slowly (days to months; type II). A nonequilibrium model conceptually similar to the Selim-Amacher model may be applicable to suspended sediment as well as bottom sediment if a kinetic model becomes warranted.

3. Sorption of Hydrophobic Organics

Sorption of nonpolar, hydrophobic organic contaminants (HOCs) appears to be primarily a solvophobic or solution phenomenon[10-12] as opposed to the site-specific adsorption of polar or ionic species. Sorption of HOCs is further complicated by HOC associations with the dissolved (DOM) and particulate (POM) organic materials present in most natural systems. Humic and fulvic acids are the DOMs of primary importance. Humics have a significant influence only on HOCs with K_{ow} greater than 10^4 according to Karickhoff et al.[12] If sorption to POM aggregates is determined to be important, diffusion-limited sorption may be operative, which would mandate a kinetic modeling framework.

POM-water partitioning of HOCs may be expressed in particulate organic carbon (POC) equivalents. The POC fraction tends to be associated with finer particle sizes (<50 μm). The organic carbon (OC) partitioning coefficient K_{oc} is related to the empirically determined, bulk sediment-water partitioning coefficient (K_p) and particulate organic carbon (POC) fraction by

$$K_{oc} = \frac{K_p}{POC}$$

The K_{oc} of HOCs appears strongly correlated with the octanol-water partitioning coefficient, K_{ow}, by the expression $K_{oc} = 0.63\,K_{ow}$ ($r^2 = 0.96$), ignoring the nonzero intercept for reasons of simplicity and data scarcity in the low $K_{ow} - K_{oc}$ region.[12] K_{ow} values are available for most common organic pollutants.

Combining the previous two relations, the bulk partitioning coefficient may be estimated by:

$$K_p = 0.63\ K_{ow}(POC) \tag{5.4}$$

Karickhoff et al.[12] observed that K_p is linear over a large range of organics concentration and "relatively independent" of sediment concentration and ionic strength effects.

Reversible DOM-water partitioning of HOCs may be described with a DOM distribution coefficient,

$$K_{DOM} = \frac{C_{DOM}}{C_D[DOM]}$$

where C_{DOM} is the concentration of HOC associated with DOM, C_D is the concentration of a true aqueous HOC solution. McCarthy and Jimenez[13] determine an empirical relationship between K_{DOM} and K_{ow} by evaluating binding of several polycyclic aromatic hydrocarbons (PAHs; benzo[a]pyrene, benzoanthracene, and anthracene) on commercially available (Aldrich) dissolved humic material:

$$\log K_{DOM} = (1.03 \log K_{ow}) - 0.5$$

B. AIR-WATER EXCHANGE

Mass transfer of organic and metal toxics at the air-water interface includes water-to-air processes, such as:

- Volatilization
- Diffusion
- Entrainment of sea-salt aerosols

Air-to-water transfer (loading) includes wet and dry deposition of contaminants from distal sources as well as locally recycled contaminants. Any surface active contaminants may accumulate in the surface *microlayer* to concentrations orders of magnitude greater than bulk concentration[14,15] affecting all of the above processes.

Air-water exchanges are particularly important for volatile organic contaminants. For example, atmospheric deposition is established as one of the most important pathways for PCB transport in the Great Lakes system.[16] The rate of volatilization for a compound is proportional to its Henry's Law constant (H) defined as H = P/C, where P is the vapor pressure and C is the dissolved concentration. The compounds molecular size, functional groups, structure, and dielectric constant influence the rate of volatilization.

Mackey et al.[17] suggested that a two-film, dual-resistance model may adequately describe the volatilization phenomenon, though they did not consider quantitatively the influence of a surface microlayer. The dual-resistance model treats the air-water interface as consisting of two, infinitesimally thin films separating the bulk air and water compartments. The air-film and water-film possess different transfer coefficients, the slower of which limits diffusive mass transfer. For example, molecular oxygen is controlled by water-film diffusion. Only the slower transfer coefficient (higher resistance) is needed for flux calculations. Volatilization may be described by

$$\frac{dC}{dt} = \frac{K_L}{d}\left[\frac{C_{atm}}{H_{eq}} - f_d C\right] \tag{5.5}$$

where

K_L = overall transfer coefficient
d = average depth
C_{atm} = atmospheric composition (typically assumed = 0)
H_{eq} = Henry's Law Constant at equilibrium
f_d = compounds dissolved fraction

See Mackey et al.,[17] Liss and Slater,[18] or Thomas[19] for a more detailed discussion of air-water exchange and the dual-resistance model.

Sea salt aerosol entrainment is a well-known phenomenon which may also be a significant influence on water-to-air transfer of both metals and organics, particularly surface active organics such as HOCs. The rate of sea salt transfer is a function of a number of physical parameters such as wind velocity, surface roughness, air and water temperatures, relative humidity. Much of the aerosol is likely to reenter the water by dry deposition or wash-out by precipitation.

Surface activity of a compound may be quantified by its Gibbs surface energy, Γ, for which positive values indicate increased surface activity, since

$$\Gamma_i = -\frac{a_i}{RT}\left[\frac{d\gamma}{da_i}\right]$$

where a_i is the activity of species i and g is surface tension.[20,21] Microlayer thickness is on the order of 100 mm, but in practice is defined operationally, with reported thicknesses ranging from 0.001 to 1000 mm. The microlayer may represent only a temporary holdup to transfer with residence times ranging from seconds to days.[14] The influence of the microlayer may prove to be an important factor in fate, transport, and bioavailability modeling, particularly if the compound is susceptible to photolysis, or if the elevated concentrations have lethal or sublethal effects on neuston communities. Gibson[22] suggests that observed differences in plankton populations within the microlayer vs. at one-meter depth may be related to microlayer toxics, though causality is not demonstrated.

Gibson,[22] Hardy et al.,[23] and researchers presently active at the Virginia Marine Science Institute (VIMS) are among the few to investigate the microlayer in Chesapeake Bay or its tributary estuaries. Hardy et al.[23] examined the microlayer at 12 sites during May 1986 and observed elevated levels of metals (Cd, Cu, Pb, Zn, etc.), alkanes, and aromatics (e.g., PAHs). Researchers at VIMS[24] are examining the microlayer phenomenon as part of a ten year study (1983 to 1993).

Quantifying air-water fluxes may be attempted with a range in degrees of sophistication. Some of the options include

1. Ignore air-water exchange in cases where such is negligible relative to other kinetic or transport processes
2. Assign constant or time-variable influx (loading) and efflux rates
3. Simple air-water partitioning based on known bulk water or air concentration and Henry's law constant (H)
4. Utilize two-layer (dual resistance) model which requires exchange rate constants and may require wind speed data for the gas film term
5. Detailed modeling including the above factors (4,3) plus consideration of microlayer phenomena, wind speed, surface roughness, and aerosol entrainment

Toxic loading rates at the air-water interface from nonpoint sources (or even distal point sources) are poorly quantitated. Dry and wet deposition data are scarce.

C. BIOCONCENTRATION

Bioconcentration is the active and passive accumulation of compounds within organism tissues to levels above ambient concentrations. Bioconcentration includes bioaccumulation which is the passive uptake of compounds through membranes (gills, cell walls) as opposed to active uptake by ingestion of contaminated organisms, sediment or water. Bioaccumulation of many organic contaminants is related to their hydrophobic (lipophilic) character, which may be correlated with the K_{ow}.

Bioconcentration may be quantified with the bioconcentration factor (BCF) which is equal to the concentration in organism divided by the concentration in water. In essence, BCF describes body loads as a simple partitioning. Increasing bioconcentration at successive trophic levels is known as biological magnification.[2]

The rate of change in concentration of a nonreactive toxicant in tissue residue, F in μg/kg may be described by

$$\frac{dF}{dt} = eK_1 \frac{C}{B} - K_2 F \tag{5.6}$$

where

e = efficiency of adsorption,
K_1 = uptake rate constant (d^{-1})
K_2 = depuration rate constant (d^{-1})
B = organism biomass (kg/l)

Internal biotransformations of reactive contaminants complicate the situation and are considered below.

Estuarine ecosystems are so diverse and complicated that a small group or organism species must be selected for bioconcentration monitoring and modeling. Benthic filter feeders such as oysters or mussels are commonly selected as monitor organisms for sediment associated contamination, as are polychaete worms and bottom-feeding fish. Benthic crustaceans such as the environmentally sensitive amphipods are often evaluated. Uptake of toxics concentrated in the surface microlayer may be monitored by analysis of select species of neuston (surface-dwelling organisms) such as larvae of oyster and other species, though no consensus exists as to the optimal species. Likewise, no particular phytoplankton or zooplankton species is consistently used to represent biomagnification at these trophic levels.

D. CHEMICAL SPECIATION

Environmental physicochemical behavior and bioavailability of toxics may be significantly affected by the aqueous form of the contaminant.[25-29] Speciation refers to the tendency for a compound or element to assume a variety of aqueous forms (e.g., valence states, complexes, or associations) in response to environmental factors such as Eh, pH, and activity of organic and inorganic, complexing legands.

Changes in speciation in response to environmental changes are typically assumed to be completed rapidly, on the order of minutes to hours. Thus speciation is generally described in terms of thermodynamics rather than kinetics. Equilibrium speciation of many common toxicants, particularly for metals in dilute aqueous solutions, may be readily predicted. Several well documented thermodynamic models are available such as WATEQ4F,[30] the latest in the WATEW series, and MINTEQ (Felmay et al.[31]) and MINTEQA2.[32]

Aqueous associations with dissolved organic carbon species, particularly humic and fulvic acids, can significantly enhance the transport of hydrophobic organic compounds. McCarthy[26] reviews the effects of DHM on HOC's with K_{ow}s greater than 10^4 and reports that bioavailability of metals or organics may be enhanced or reduced, depending on the organism. The organic chemistry of natural waters is commonly complex and poorly defined. Thermodynamic data may not be available to describe the interactions of these organic ligands and toxic or other species, rendering the results of any of the existing equilibrium models equivocal at best.[27]

Consideration of chemical speciation becomes essential if a nonequilibrium system is adopted for any of the key chemical reactions modeled, since reaction pathways require information on the distribution of reactant species.[5] Few kinetic speciation models have been developed. Fontaine[33] describes the kinetic framework for speciation and adsorption utilized in the model NONEQUI for metals transport in low pH streams. Fontaine[33] considers adsorption only to organics, clays, silt, and sand substrates; inorganic ion pair formation is minimal in low pH, low ionic strength waters and are not considered in the model. Estuarine water pHs tend to be neutral to slightly alkaline (6.5 to 8) and ionic strength increases dramatically with proximity to marine waters. Thus, metal speciation within an estuarine system would be decidely more complex than modeled by Fontaine,[33] probably requiring consideration of ion pairs.

Chelation with macromolecular organics tends to reduce bioavailability of metals as well,[26,34] though data are more equivocal than for organics due to additional complicating factors.

E. BIOTRANSFORMATION

Biotransformation of toxics includes both the alteration of the compound to another compound that may be either more or less toxic, and the complete biodegradation of the contaminant (e.g., mineralization). As two examples: (1) the pesticide DDT may be altered to DDE and a series of other metabolites which may be equally hazardous;[35] and (2) the methylation of relatively nonbioavailable elemental mercury to highly toxic methylmercury.

The organisms affecting the transformations produce enzymes which catalyze hydrolysis, oxidation/reduction, and acid/base reactions. Most environmentally significant biotransformations are microbially mediated, though higher organisms may metabolize certain toxics as well. The microbial reactions may be intracellular or extracellular. Metabolism by higher organisms is important in biomagnification and toxicological investigations.

Organic contaminants may serve as a carbon and energy source for heterotrophic bacteria. A recalcitrant organic contaminant may be utilized as a secondary or incidental carbon source (cometabolism) in the presence of a more thermodynamically favorable carbon source. Principal processes involved in microbial transformation are depolymerization, dehalogenation, and aromatic ring cleavage.[36] Where the organic compound is a primary carbon source, the biotransformation may follow a Michaelis-Menton type kinetic structure. The Michaelis-Menton kinetics equation is

$$\frac{dC}{dt} = -K_b C \tag{5.7}$$

where

$$K_b = \frac{X}{Y(K_m + C)}$$

μ = maximum growth rate
X = biomass concentration
Y = yield coefficient
K_m = half-saturation constant

Biotransformation is also affected by environmental factors as temperature, the availability of other nutrients. The microbial population typically requires some time to evolve enzymes capable of transforming new xenobiotics, so the degree of microbial acclimation to the contaminant becomes important in certain biotransformation.

F. PHOTOLYSIS

Photolysis is the chemical alteration of a compound due to the direct or indirect effects of light energy. Light energy sorbed by the molecule may be reradiated as heat, fluorescence or phosphorescence, or, if the energy is sufficient (quantum), the molecular structure reorganizes to accommodate the increased energy condition. Direct photolysis requires input of light energy to proceed, whereas, indirect photolysis involves an initial photolytic reaction, the product of which (e.g., ozone, peroxide) reacts with the compound of interest. The product of the initial phototransformation is known as the nontarget intermediary.

Photochemical alterations of most toxics remains poorly quantified. Photolysis may reduce the compounds toxicity as in the case of the indirect photoreduction of the highly toxic and carcinogenic Cr(VI) to the innocuous Cr(III) by indirect reaction with a hydrogen peroxide intermediary.[37] Photolysis may also increase toxicity as in the case of the indirect photo-oxidation of the insecticide parathion to paraoxon by reaction with peroxides.[37]

The rate of direct photolysis may be described by

$$\frac{dC}{dt} = -K_a \phi C \tag{5.8}$$

where

C = concentration
K_a = rate constant for light adsorption
ϕ = quantum yield of reaction (moles toxics reacted/einsteins adsorbed)

The ϕ may be considered as the efficiency of the phototransformation. The kinetics of indirect photolysis is described by

$$\frac{dC}{dt} = -K_2 CX = -K_p C \tag{5.9}$$

where

K_2 = rate constant for indirect photolysis
X = concentration of nontarget intermediary
K_p = pseudo first order rate constant (= K_2X)

G. HYDROLYSIS

Certain organic compounds, susceptible to nucleophilic attack, tend to hydrolyze in reactions with H_2O, HO^-, or H_3O^+. Acid-base and hydration reactions are not considered hydrolysis. Common susceptible organics include esters, amides, amines, carbamates, alkyl halides, and epoxides.

The general rate expression for hydrolysis of neutral organic compounds in pure water is[38]

$$\frac{dC}{dt} = -K_a [H^+][C] - K_n [C] - K_b [HO^-][C] \tag{5.10}$$

where

C = toxicant concentration
K_a, K_n, K_b = rate constants for acid-, neutral-, and base-catalyzed reactions, respectively
[ ] = molar concentrations

Acid- and base-catalyzed hydrolysis reactions are obviously pH-dependent. Hydrolysis may also be catalyzed by metal ions or ionic strength effects.[38] The kinetics may be first order or second order. Hydrolysis of adsorbed contaminants is possible though the reaction rate is usually slower.[38]

H. DISSOLUTION AND PRECIPITATION

The dissolution or precipitation of solids phases involves mass transfer between aqueous and solid phases (exclusive of adsorption phenomena). The toxicant may be involved directly or passively in the reaction. The solid phase may serve as

1. A source, if the toxic species is a significant component in a soluble solid phase (e.g., redox- or pH-sensitive or chemically unstable)
2. A sink, if the toxic species is a key structural element or is trapped passively in a precipitation reaction

For example, hydrous metal oxides may precipitate in response to appropriate increases in pH or Eh condition, incorporating toxic metal species. The precipitation of sulfide minerals under anaerobic conditions in anoxic bottom waters or, more commonly, in deep sediments, may serve as an important sink for metal species (Cu, Zn, Fe, Pb, Hg, etc.). Equilibrium mass transfer models such as PHREEQE[39] are available, which consider acid-base and redox reactions for a limited suite of metals and solids species.

I. SEDIMENT DYNAMICS, DIAGENESIS, AND DIFFUSIVE EXCHANGE

An accurate treatment of sediment dynamics is essential to transport and bioavailability modeling of sediment-associated toxics, such as metal cations and highly hydrophobic organics. Processes to be considered in such modeling include:

1. Accurate prediction of hydrodynamic transport (advection-dispersion) and settling of suspended solids in the water column, complicated in estuarine systems by stratified flow, salinity gradients, and tidal influences
2. Particulate exchange between water column and bed sediment compartments, with consideration of important hydrodynamic and physicochemical controls on deposition and remobilization
3. Sediment diagenetic processes such as organic carbon metabolism (methanogenesis, denitrification, sulfate reduction), metal methylation, and sulfide formation
4. Diffusive exchange between the water column and bed sediment compartments, particularly oxygen (SOD) and nutrient fluxes

All of these processes may affect directly or indirectly the environmental behavior of contaminants. Spatial variability and incomplete understanding of controls on these processes complicates the modeling effort, often requiring considerable site-specific field measurement and laboratory experimentation.

All of the kinetic phenomena discussed in the previous section may be active in the sediment layer, though air-water exchange is replaced by water-bed exchange and direct photolysis rapidly becomes insignificant with increasing water depth. The physical, chemical, and biological environment in the sediment column contrasts sharply with the overlying water column, often requiring a separate set of equilibrium or kinetic coefficients. Particulates concentration, temperature, Eh-pH conditions, and the microbial population diversity and density are likely to be different in the sediment than in the water column.

1. Sediment Transport Dynamics

Estuarine sediments constitute a heterogeneous mixture of inorganic particulate sizes (sand, silt, clay) and mineralogy (tectosilicates, phyllosilicates, oxides) as well as organic detritus. These sediments tend to accumulate over a long time scale (years), though over shorter time frames (tidal cycle) they may be quite dynamic. The rate of sediment accumulation is a function of net sediment budget, which is controlled by climatic, physiographic, and anthropogenic factors within the drainage basin. Channel deepening by dredging to accommodate increased shipping traffic tends to accelerate the rate of deposition relative to the natural equilibrium rates, as has been observed in the Hampton Roads area at the confluence of the James and Elizabeth Rivers in Virginia.

Fine-grained, cohesive sediments are particularly important as substrates for sediment-associated toxics due to their larger surface area to volume ratio and the high cation exchange capacity of clay minerals. Mehta et al.[40,41] have summarized the salient aspects of the physical transport of cohesive sediments.

Many of the important processes in cohesive sediment transport are the same as for noncohesive sediment: settling, deposition, consolidation, resuspension, erosion, and lateral transport (siltation), all largely controlled by the sediment response to hydrodynamic forces. Cohesive sediment dynamics are complicated further by biologic and physicochemical processes such as flocculation and dispersion, filtration and bioaggregation, and reentrainment. The controls on hydrodynamic behavior of high density sediment suspensions and consolidating cohesive sediments is poorly understood.[40]

Settling velocities in the water column are influenced by the size, shape, relative density, and concentration of the particulates. The slow settling velocity of dispersed fine sediments, with which many toxics are associated, suggests the critical importance of accurate hydrodynamic modeling as part of toxics transport and fate modeling.[41] Settling velocities of cohesive sediments may be increased by aggregation into larger particles that may occur abiotically in response to increasing ionic strength (salinity) causing flocculation *or* by the formation of fecal pellets by filter feeders in the water column (e.g., copepods) and benthos (e.g., mollusks). Benthic macrophytes may act as sediment traps by retarding shear stresses at the sediment-water interface, thereby enhancing local sediment accumulation.

Deposition occurs when gravitational forces exceed turbulent forces and particles reach the sediment-water interface. Shear stress due to water advection must subside below the critical shear stress for the sediment. For example, the turbidity maximum zone is an area of considerable turbulence due to the convergence of flows from opposing directions, preventing deposition. The nature of deposition and bed characteristics is a product of the hydrodynamic and biogeochemical environment. The hydrodynamic conditions, with additional physicochemical influences, control the grain size sorting within deposits and bed morphology. Bed characteristics in turn influence erodability and diffusive exchange rates.

Consolidation is an important site-specific process that increases bed density and viscosity (non-Newtonian rheology) and thereby reduces erodability. The process proceeds in response to continued particulate loading and dewatering of sediment interstices.

Mehta et al.[40] support the classification of sediment into four groups: (1) horizontally mobile suspensions, (2) horizontally stationary, high density suspensions, (3) consolidating deposits, and (4) consolidated bed sediments. They suggest that the transfer of deposited solids from the bed to the water column be classified into at least three categories: *reentrainment* of stationary suspensions, *resuspension* of partially consolidated sediments, and *erosion* of consolidated sediments. For estuaries, they advocate a dual dynamic system: an upper layer as a suspension which undergoes alternating deposition during slack times and reentrainment or resuspension with increased flow velocities during peak ebb and flood stages. The lower sediment layer is relatively stable, subject to occasional scour.

2. Sediment Diagenesis and Diffusive Exchange

Diagenesis is the net effect of all physical and biogeochemical processes active in the sediment layer. Important chemical diagenetic processes are largely microbially mediated, such as the consumption of oxygen and nutrients by aerobic and anaerobic, heterotrophic bacteria. Depletion of these constituents, creates increasingly anaerobic conditions (more negative Eh) with depth in the sediment. Altering the Eh-pH conditions affects contaminant speciation and solubility and the solubility of the substrates (particularly Fe and Mn oxides). Under anaerobic conditions, nitrate- and sulfate-reducing bacteria activity may further affect metal speciation and bioavailability. If sufficient sulfate is available, as would be the case in marine and estuarine waters, metal sulfides may be generated, removing toxic metals.

Geochemical alterations induced by sediment microbial activity or repartitioning of sediment-associated toxics settling from the water column may create concentration gradients for particular toxics, favoring flux into the water column. Flux of dissolved oxygen into the sediment is critical to the decomposition of organic carbon by aerobic bacteria.

III. SIMPLIFIED FATE AND TRANSPORT ANALYSIS OF CHEMICALS

The complexity of the transport and transformation processes which influence the fate of toxicants require additional analytical tools beyond those required for conventional pollutants such as BOD and nutrients. Construction of these tools require the formulation of the kinetics and derivation of the associated kinetic coefficients. Although most kinetic formulations may be parameterized by first-order reactions, derivation of the coefficients needs a significant amount of data. It is not always possible to independently derive all the kinetic coefficients. It is therefore, important to identify the key process(es) related to the fate of the pollutant. In particular, for the over 100 organic priority pollutants, sorption processes appear to be the most important for the majority of the priority pollutants.[42] Further, different levels of complexity are usually considered in assessing the aquatic fate of a pollutant:

- Treating the pollutant as a conservative substance
- Considering transport and speciation processes
- Considering transformation, transport, and speciation processes

Each option has advantages and limitations. In the following paragraphs, a simplified approach to modeling the fate of chemicals is presented.

Lung et al.[43] conducted a study of calculating the total chemical concentrations for consumer product chemicals in a number of major estuaries in the U.S. using a simplified mass transport and kinetics model. Estuaries throughout the United States receive chemical inputs from a variety of sources. Among the sources, publicly owned treatment works (POTW) discharges into estuaries are a major source of consumer chemicals (e.g., detergents). The fate and transport of these chemicals in a receiving estuary is an important aspect in the overall exposure analysis in the environment. Processes affecting the fate and transport of consumer product chemicals include mass transport, biodegradation, photolysis, volatilization, sorption, and subsequent settling/sedimentation.

A. MODELING APPROACH

To predict the fate and transport of the chemicals and to quantify the chemical exposure levels in estuaries, a modeling framework has been developed in which several of these processes are incorporated with physical characteristics such as estuarine circulation and mass transport. In general, usage rates of consumer product chemicals were relatively constant and supported by over ten years of influent wastewater monitoring data, suggesting that tidally averaged transport patterns using a steady-state fate-transport solution are appropriate for examining the exposure levels of these chemicals. Depending on the seasonal nature of freshwater flow, the mass transport pattern may be one-dimensional (longitudinal-vertical) in many partially mixed estuaries. In addition to mass transport, loss of chemical from the water column (e.g., by biodegradation) is accounted for by a first-order kinetics algorithm. BOD data (both point source and receiving water data) are used to validate the modeling framework.

Wastewater flow and treatment process (primary, trickling filter, activated sludge, or lagoon) data were obtained for the POTWs which discharge to the estuary. Wastewater treatment plant chemical removal extents that were determined during specific monitoring programs or predicted from laboratory tests were used to calculate effluent loads at the point of discharge to

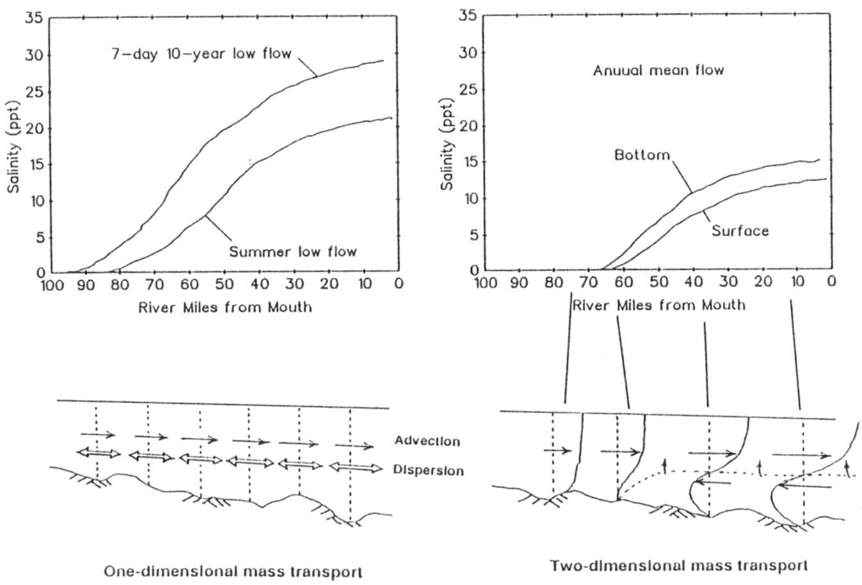

FIGURE 1. Mass transport patterns and salinity profiles for different freshwater flows.

each estuary. Removal extents were calculated as the percent removal between influent and final effluent wastewater. These data were used in the modeling framework to predict chemical concentrations in the estuaries under various freshwater flow conditions.

As a first step analysis, chemical concentrations are calculated on a tidally-averaged basis. Tidally-averaged salinity distributions and estuarine mass transport patterns are strongly influenced by freshwater flows in the estuarine system. As shown in Figure 1, lower freshwater flows such as seven-day ten-year low flow and summer low flow would normally cause significant salinity intrusion inland, resulting in better mixing in the water column with little vertical salinity stratification. As a result, one-dimensional (longitudinal) mass transport models provide a good approximation of the physical process. On the other hand, higher freshwater flow (annual mean flow) would tend to deliver freshwater to the surface layer of the water column with salinity intrusion in the bottom layer, thereby generating a moderate salinity stratification (Figure 1). This is a typical tidally-averaged circulation and salt balance pattern observed in many partially mixing estuaries (Figure 1). In this case, a two-layer transport pattern is appropriate to characterize the mass transport. In the one-dimensional models, mass transport is characterized by longitudinal advection and dispersion coefficients. In the two-dimensional (two-layer) models, the circulation pattern is computed using the methodology by Lung[44] and Lung and O'Connor[45] (see Chapter 3).

Sorption of chemicals can be quantitatively expressed by a sediment-water partition coefficient.[46] Thus, suspended solids concentrations in the water column and the sediment need to be quantified in the modeling analysis to account for sediment-water interactions. Estuarine bed movements of sediments are another important factor to be incorporated into the analysis. As a result, a comprehensive modeling framework is required to address the fate and transport of these chemicals in an estuarine system.[2] However, for chemicals which are not readily adsorbed, other processes such as biodegradation and photolysis may be dominating in the water column. To simplify the analysis, a first-order term for the net removal (loss) of the chemicals from the water column is formulated to characterize the disappearance of the chemicals from the water column caused by biodegradation of dissolved and particulate (sorbed) components and sedimentation of the sorbed component (see Thomann and Mueller[1]). The modeling framework uses a net settling rate that includes particle settling with sorbed chemical loss to the sediment and the opposing resuspension of particles (with sorbed chemical) back into the water column. The settling process generally dominates resuspension. Similarly, volatilization and uptake of dissolved chemicals by biota are also envisioned as part of the net disappearance process. Such an approximation is particularly valid for chemicals that are biodegradable.

B. CALCULATING TOTAL CHEMICAL CONCENTRATIONS

In this analysis, linear alkylbenzene sulfonate (LAS), a detergent builder, which is readily biodegradable and nonvolatile, is chosen as the first chemical to be modeled. For the LAS scenarios considered, water column concentrations are based on instream biodegradation only (i.e., no other removal process is considered), thus the predicted concentrations are conservative.

Based on an extensive monitoring database, the influent concentration of LAS at POTWs is conservatively assumed to be 5 mg/l.[47-49] Depending on the treatment level, the total chemical concentration in the POTW effluent is calculated. The overall decay rate for LAS in the water column is assumed to be 0.09 d^{-1}, based on reported biodegradation half-lives of six to nine days.[50,51] Ambient concentrations of LAS in the water column (i.e., locations remote from wastewater discharges) conservatively assigned a value of 0.1 μg/l.

The modeling framework has been applied to 17 major estuaries in the U.S.[43] These estuaries are selected because they are located with large population centers which have significant consumer product consumption. Only results from select estuarine systems are presented in this section. Figure 2 shows the results of salinity, BOD_5, and LAS for the Delaware Estuary under the summer low flow condition. The one-dimensional mass transport pattern is determined using salinity as a conservative tracer. The discrepancies between the model results and the data for the Delaware Estuaries are equivalent to the significant amount of algal biomass in the water column. Next, the simulated LAS concentration profile parallels the BOD simulation except that calculated LAS concentrations are two or three orders of magnitude lower than BOD. Such a similarity is expected because the biodegradation rates for LAS and BOD are in the same order of magnitude and substantiates the validity of the LAS simulations.

Figure 3 shows the calculated LAS concentrations (in μg/l) in San Francisco Bay using a two-dimensional model under a steady-state condition.[43] The mass transport coefficients were derived from a study by Selleck et al.[52] High LAS concentrations are located in the southern tip of the bay, reflecting significant population in that area. Another example of the model application to LAS calculations is Tampa Bay, Florida[43] (Figure 4) using mass transport coefficients developed in a number of studies.[53-55] Both models are run using the framework HAR03 under steady-state conditions. In both cases, two-dimensional salinity distributions were matched by the model results prior to the LAS calculations.[43]

The modeling framework was also applied in a two-dimensional (longitudinal-vertical) configuration to the James Estuary.[43] The LAS results are presented in Figure 5. The LAS concentrations in the surface layer are slightly higher than those in the bottom layer because the surface layer receives the POTW discharges. The two peak LAS concentrations in the surface layer are due to the POTW discharges in the Richmond and Hopewell areas, respectively. Also shown in Figure 5 are the salinity profiles in both layer matching the data from the model calibration analysis of mass transport.[43]

C. CALCULATING DISSOLVED AND PARTICULATE CONCENTRATIONS

The above example of the James Estuary shows the fate of total chemical concentrations in the water column. When considering bioavailability of the pollutant, it becomes necessary to partition the total chemical concentration into particulate and dissolved phases. To calculate the dissolved and particulate concentrations, one needs to determine the suspended solids concentrations in the estuary. The methodology presented in Chapter 3 on quantifying the suspended solids concentrations in two-dimensional distributions would be used. The mass balance equation for the suspended solids concentration, m, is

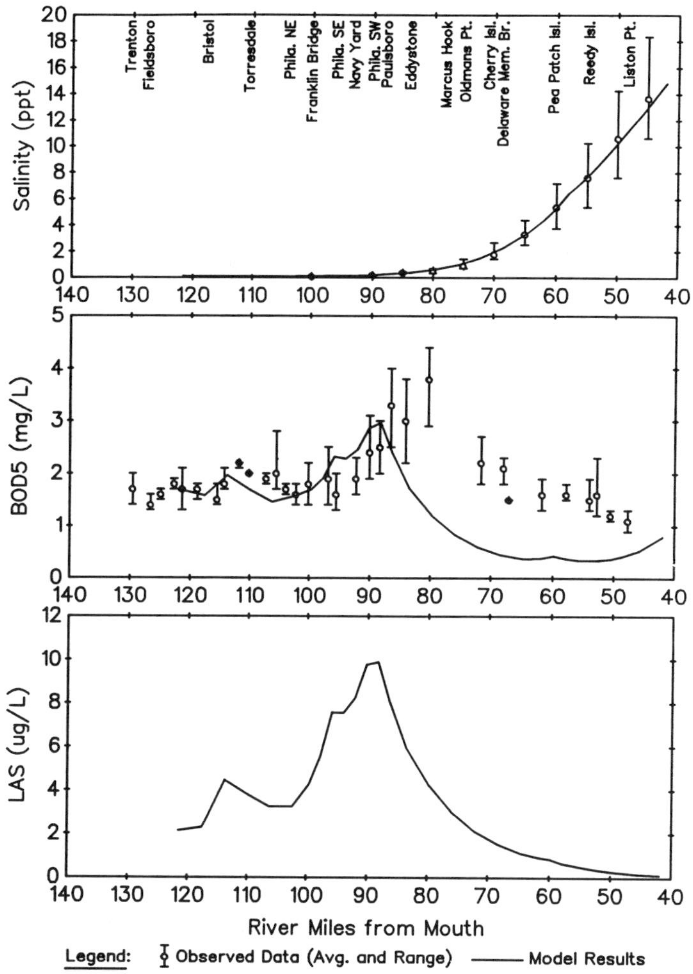

FIGURE 2. Salinity, BOD_5, and LAS model calculations for the Delaware Estuary.

$$\frac{\partial}{\partial x}(um)+\frac{\partial}{\partial y}[(v+v_s)m]=\frac{\partial}{\partial y}\left(\xi\frac{\partial m}{\partial y}\right)$$

The settling velocity of solids, v_s, is added to the vertical velocity to account for the settling of suspended sediment through the water column.

The dissolved and sorbed concentrations of chemicals vary as a function of the suspended solids concentrations and the partition coefficient, K_p. The linear Langmuir isotherm, shown in Equation 5.1, is used for the calculations and local instantaneous equilibrium of the chemical is also assumed. There has been research indicating that the value of the partition coefficient is independent of the suspended solids concentrations, though related to the organic content of the solids.[8] For these calculations the partition coefficient is assumed constant throughout the estuary.

The sorption phenomena can be characterized by incorporating appropriate terms into the mass balance equation. The dissolved and sorbed concentrations of the chemical can be treated with separate mass balance equations. Under steady-state, tidally-averaged conditions, the equation for the dissolved concentration, C, is

$$u\frac{\partial C}{\partial x}+v\frac{\partial C}{\partial y}-\frac{\partial}{\partial y}\left(\xi\frac{\partial C}{\partial y}\right)-KC-K_1mC+K_2p=0 \tag{5.11}$$

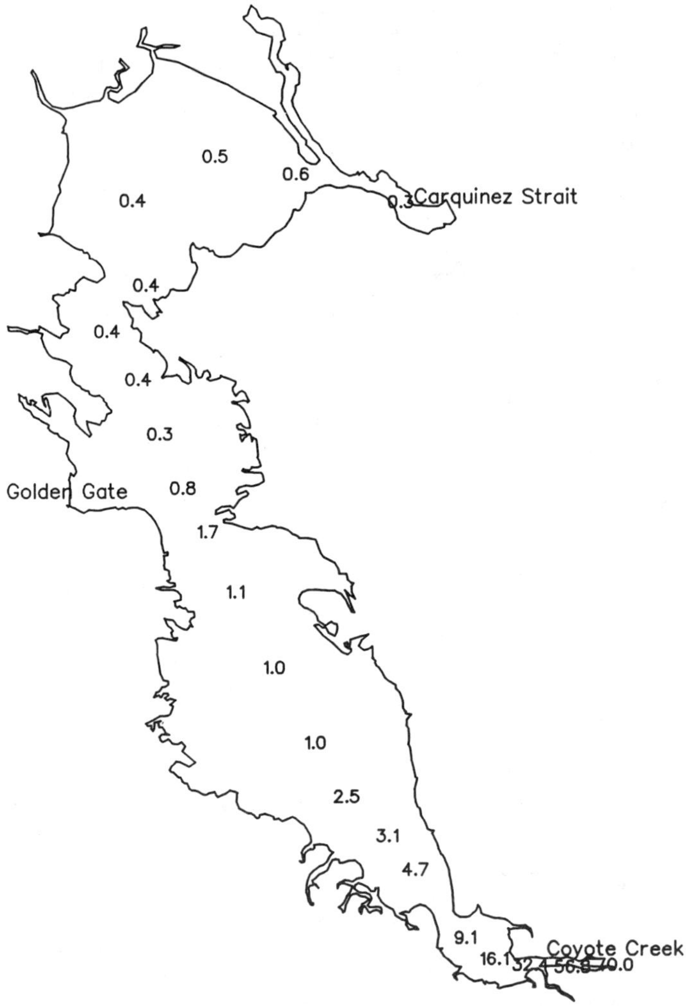

FIGURE 3. LAS model calculations for San Francisco Bay.

$$u\frac{\partial p}{\partial x}+(v-v_s)\frac{\partial p}{\partial y}-\frac{\partial}{\partial y}\left(\xi\frac{\partial p}{\partial y}\right)+K_1mC-K_2p=0 \tag{5.12}$$

where

m = suspended solids concentrations
K = first-order kinetic loss rate (d^{-1})
K_1 = adsorption rate coefficient
K_2 = desorption rate coefficient

The last two terms in each of the equations represent adsorption and desorption, respectively. The adsorption of dissolved contaminant onto the suspended particles results in an increased concentration of particulate chemical phase, while the reverse contributes to an increase dissolved phase concentration and corresponding decrease in particulate phase.

The sorbed concentrations is related to the suspended solids concentration as

$$p = rm$$

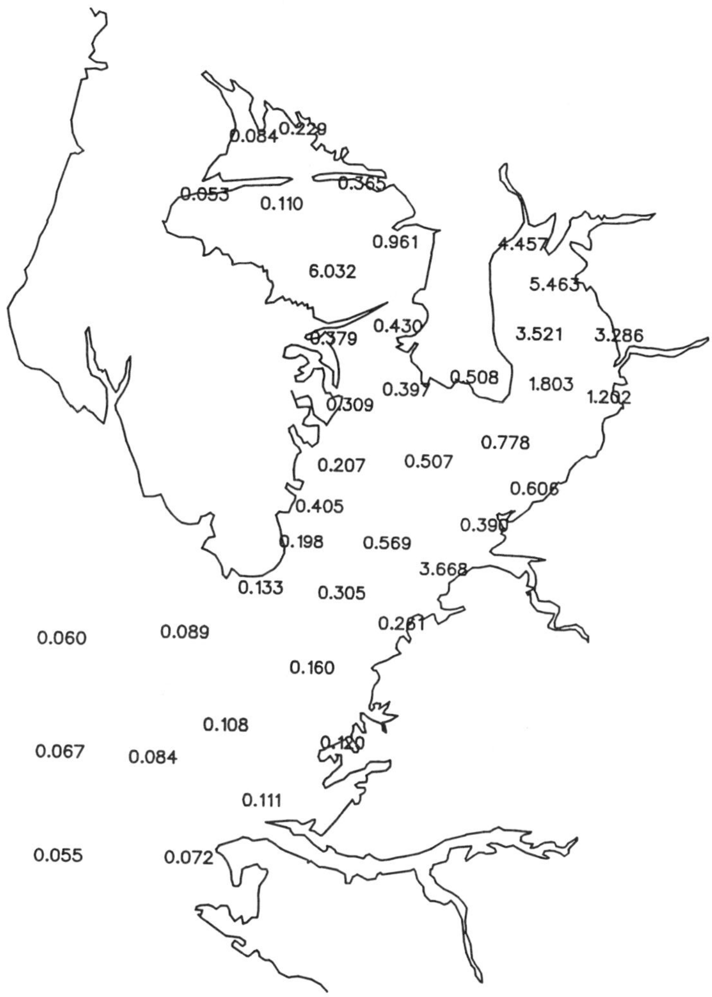

FIGURE 4. LAS model calculations for Tampa Bay.

where r represents the sorbed phase concentration of the chemical in μg/l per mg/l. Following the linear adsorption isotherm, the partition coefficient, K_p, is expressed as

$$K_p = \frac{r}{C}$$

Upon substitution, the total chemical concentration can be written in terms of the dissolved chemical concentration as

$$C_t = C + K_p mC$$

The fraction of dissolved chemical concentration, f_d, can be expressed as

$$f_d = \frac{C}{C_t} = \frac{1}{1 + K_p m}$$

The dissolved phase concentration is then

$$C = \frac{C_t}{1 + K_p m}$$

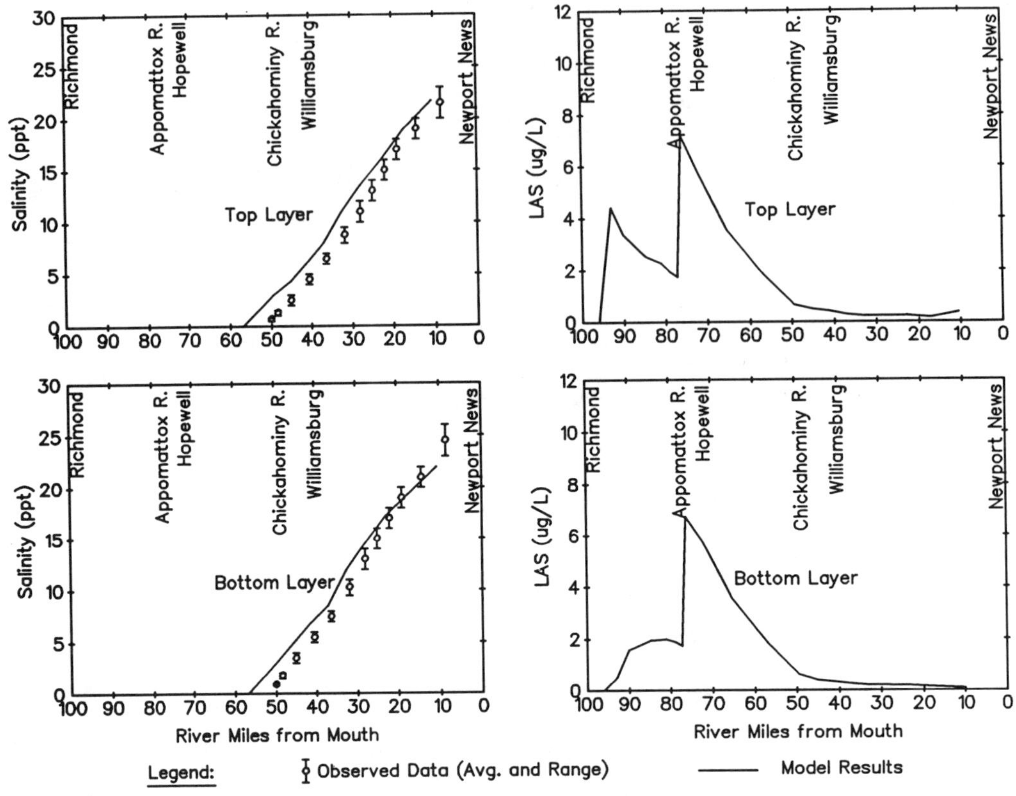

FIGURE 5. Salinity, BOD_5, and LAS model calculations for the James Estuary.

Likewise, the particulate fraction of the total chemical concentration, f_p, is

$$f_p = \frac{p}{C_t} = \frac{K_p m}{1 + K_p m}$$

and the particulate phase concentration is

$$p = \frac{K_p m C_t}{1 + K_p m}$$

These relationships are used to predict the dissolved and particulate phase concentrations, given the partition coefficient, after the total chemical concentration and suspended solids concentration have been computed. Combining Equations 5.11 and 5.12 results in a single equation to describe the chemical transport:

$$u\frac{\partial C_t}{\partial x} + v\frac{\partial C_t}{\partial y} - v_s\frac{\partial p}{\partial y} - \frac{\partial}{\partial y}\left(\xi\frac{\partial C_t}{\partial y}\right) - KC = 0 \tag{5.13}$$

An assumption implicit to Equation (5.13) is that of instantaneous equilibrium. Since the adsorption-desorption rate is so much greater than the other mass transfer and kinetic rates, it can be assumed that the sorbed and dissolved chemical phases are in equilibrium with respect to the rest of the system.[46] Thus, an equilibrium is maintained, defined by the partition coefficient, and the sorption process does not need to be included in the equation. O'Connor and

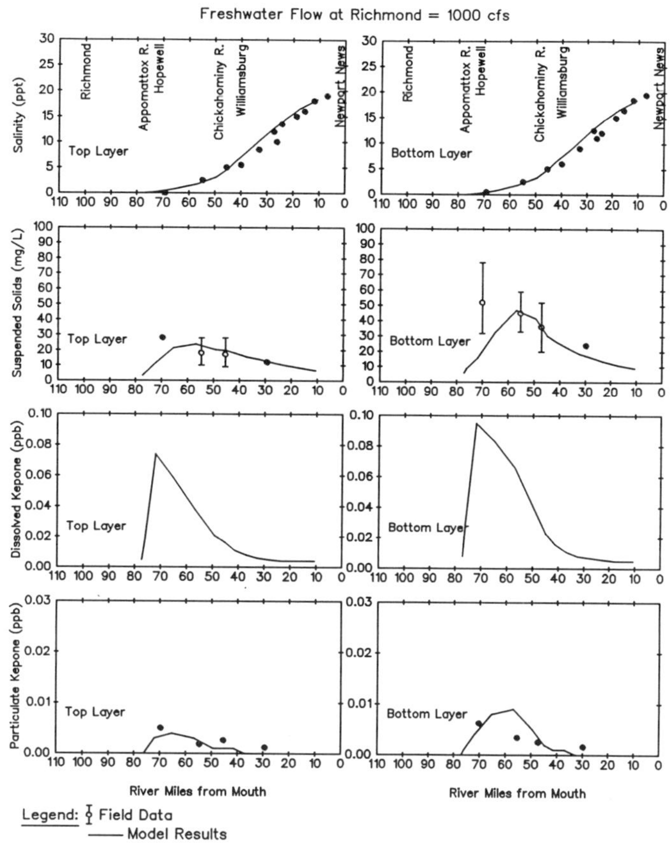

FIGURE 6. Kepone model calculations for the James Estuary.

Farley[56] performed an analysis of the James Estuary that justified the assumption of instantaneous equilibrium.

The assumption of instantaneous equilibrium, with substitution of the total concentration for the particulate fraction, gives the complete form of the steady-state mass balance equation:

$$u\frac{\partial C_t}{\partial x} + v\frac{\partial C_t}{\partial y} - v_s\frac{\partial}{\partial y}(f_pC_t) - \xi\frac{\partial^2 C_t}{\partial y^2} - Kf_dC_t + (\text{sources} - \text{sinks}) = 0 \qquad (5.14)$$

The modeling framework was applied to the James Estuary (Figure 6) to calculate Kepone concentrations in the water column.[56] Discharges of Kepone from the Hopewell area, which began as early as 1967, continued to enter the estuarine system until 1973. Due to the extremely slow rates of volatilization and biodegradation of Kepone, these transfer routes are not important.[56] Other factors, such as the bioaccumulation in the aquatic food chain, although a most relevant parameter, may be considered insignificant in the first analysis, since the total mass of Kepone in this phase is one or two orders of magnitude less than the mass of Kepone associated with the suspended and bed sediment. As such, the equations relating to the food chain may be decoupled from those defining conditions in the suspension and in the bed, as a first approximation. The significant factors which must be taken into account are the transport of the water and the bed, the exchange between these phases and within each phase, and the interaction between the dissolved and particulate components.

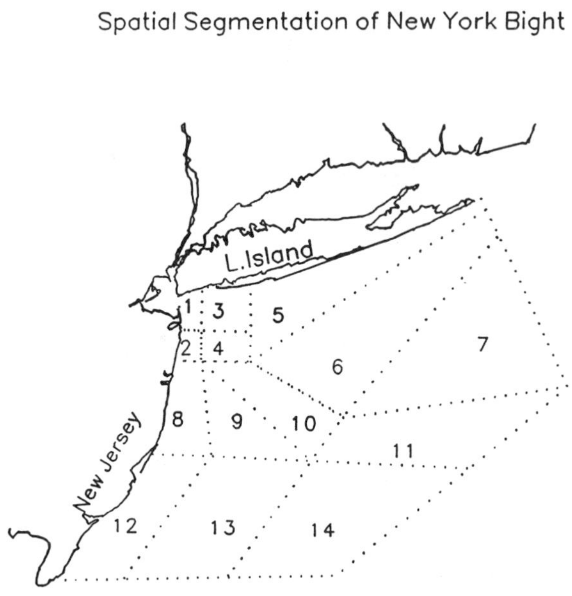

FIGURE 7. New York Bight.

Further approximation by Ruppe and Lung[57] is incorporating a net settling term of suspended solids to account for the sediment-water interactions such as resuspension and deposition. Thus, no sediment system is considered, a simplified approach compared with that presented by O'Connor et al.[58] Model results by Ruppe and Lung[57] and field data for salinity, suspended solids, and Kepone concentrations (dissolved and particulate) for the James Estuary under a freshwater flow of 1000 *cfs* are summarized in Figure 6. The settling velocity of suspended solids in the water column is 4.0 ft/d. A partition coefficient, K_p, of 3000 l/kg and a biodegradation rate of 0.008 d^{-1} in the water column were used in the modeling analysis.[43] Mass transport coefficients for the two-layer model were determined by matching the salinity distributions (Figure 6). Subsequent total suspended solids analysis determined the settling velocity. The calculated Kepone concentrations sorbed on the suspended solids also follow the spatial trend along the estuary. Unlike the approach taken by O'Connor et al.[58] who considered the sediment-water interactions for Kepone in the James Estuary, the analysis presented in this example represents a simplified method to calculate dissolved and sorbed chemical concentrations in the water column.

IV. MODELING METALS TRANSPORT AND SPECIATION

A. NEW YORK BIGHT

The New York Bight has been receiving waste disposals, dredged spoils, and acid dumps for the past 30 to 40 years, accumulating trace metals such as Cu(II) in the water and sediment. The boundaries of the New York Bight on the western side extend from Sandy Hook to Cape May along the New Jersey coastline and the northern boundary extend from Rockaway Point to Montauk Point along the Long Island coastline. The southern boundary extends from Cape May to the shelf break at 200-m isobath and the eastern boundary is from Montauk Point to the 200-m isobath. Figure 7 shows the location and the boundaries of New York Bight.

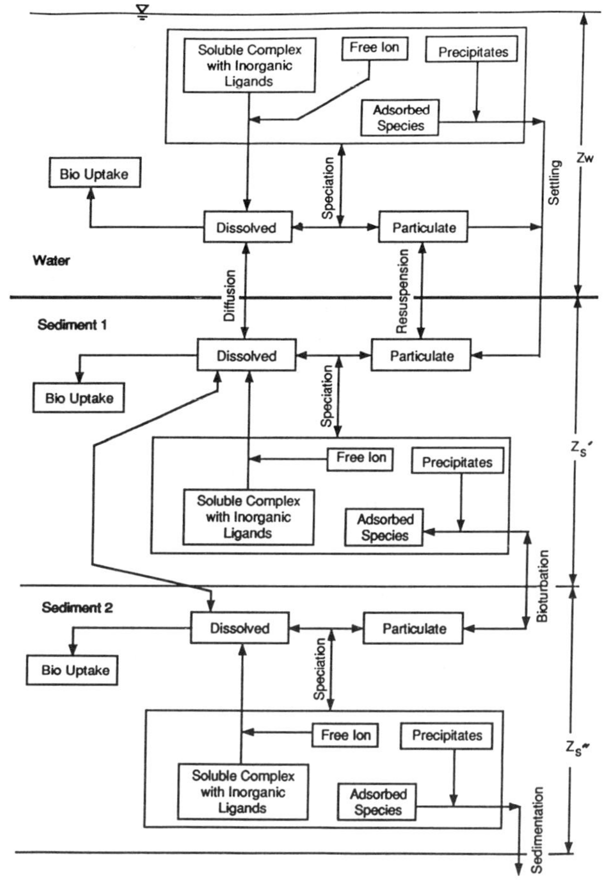

FIGURE 8. Kinetic structure for copper (II) model of New York Bight.

B. MODELING APPROACH

A three-dimensional multilayered time variable model was developed to assess the temporal and spatial distribution of dissolved and particulate Cu(II) concentrations in the water and sediment layers.[59] A chemical submodel, MINTEQA2,[32] was used to compute the equilibrium Cu(II) species concentrations. Again, a simplified box-type configuration was adopted to characterize the physical transport and kinetic processes of Cu(II) in the system. The water column is divided into two layers to represent the seasonal stratification, a phenomenon which is particularly evident at the bight apex where less dense water from the Hudson-Raritan estuary flows outward through the upper layer and denser bight water flows through the lower layer. The sediment bed is also divided into two layers; very thin top layer, which is considered to be aerobic, and a thicker bottom layer, assumed to be anaerobic. Each layer is further divided into 14 segments (Figure 7), making a total of 56 segments.

The kinetic and transport processes for dissolved and particulate Cu(II) in the water column are advection, dispersion, settling, bio-uptake, and adsorption. In the sediment column additional kinetic processes were considered namely, bioturbation causing a mixing between top and bottom sediment layers and sediment burial in the bottom sediment layer. The chemical processes under consideration are speciation and precipitation. The schematic representation of the processes involved in Cu(II) distribution is shown in Figure 8.

To focus on the sediment-water interaction the interstitial water of the thin aerobic sediment layer is considered to have diffusive exchange of dissolved Cu(II) with both the overlying water column and the underlying anaerobic sediment layer. This exchange was expressed in terms of a mass transfer coefficient which is a function of interstitial water diffusion coefficient and the characteristic length of the gradient of the dissolved Cu(II) concentration. The numerical product of these two provides the diffusive flux into or out of the sediment column. The particulate exchange in the water column is a function of advective and dispersive transport and the loss due to settling. The top sediment layer was assumed to have no accumulation or loss of particulates. However, the bottom layer of sediment was assumed to receive the settling solids particulate Cu(II) associated with them, from the water column and suffer loss of the same due to particles leaving the active sediment layer through sedimentation. The sediment surface was considered as the point of reference thus the increase in depth due to deposition of solids is a loss of mass due to burial from the active sediment layer. The process of bioturbation mixing caused by benthic microorganisms was represented by the rate of mixing of sediment and was affixed by estimating the apparent particle diffusion coefficient. Experiments relating the dependency of particle mixing caused by benthic microorganism and overlying water oxygen concentration have shown that mixing rate decrease with the decreasing dissolved oxygen concentration in the overlying water column.[58] In this study, the benthic biomass was assumed to be proportional to the labile carbon in the system thus facilitating the computation of the mixing rate.

The adsorption-desorption of Cu(II) was handled by using different partitioning coefficients from the water and sediment columns and are considered constant at 43,000 l/kg in water and 6000 l/kg in aerobic sediment.[59] The local equilibrium assumption was the basis for partitioning Cu(II) into dissolved and particulate fractions.

C. MINTEQA2 MODEL CALCULATIONS

As mentioned earlier, the chemical reactions involved in speciation and precipitation are quantified using the MINTEQA2 model.[32] In a chemical equilibrium model, the equilibrium condition for a given set of conditions and for a given set of reactants is reached by either minimizing the system free energy under mass balance constraints or by simultaneously solving the nonlinear mass action expressions and linear mass balance relationship. MINTEQA2 uses the later approach commonly known as *equilibrium constant method.* In this method, all the possible mass action constraints for reactions involving the reactants and the products are combined with the mass balance expressions for each component. The thermodynamic information required for each component involved in the reactions and also for the reactions themselves are stored in a number of databases.

Although Cu(II) reacts with a number of species in the aquatic system, it is possible to make relatively accurate predictions on its chemical speciation when only two parameters are known namely, suspended solids and total copper concentrations.[62] Furthermore, the computational facilities on adsorption-desorption provided in MINTEQA2 was not used as the data required for this option are not available. Instead, adsorption-desorption of Cu(II) was formulated using a local equilibrium assumption. Such an approach was used in other water quality modeling studies by Lung[63] and by Schnoor;[64] and analytically proven to be valid by O'Connor.[46]

D. MODEL APPLICATION

To apply this modeling framework, seasonal circulation patterns are used. The annual mean current drift over the shelf is in southwesterly direction parallel to the shore. This is usually caused by the influx of river runoff from the coast and is influenced by the cross-shelf density gradient. The general southwesterly circulation is often reversed for a period of about three

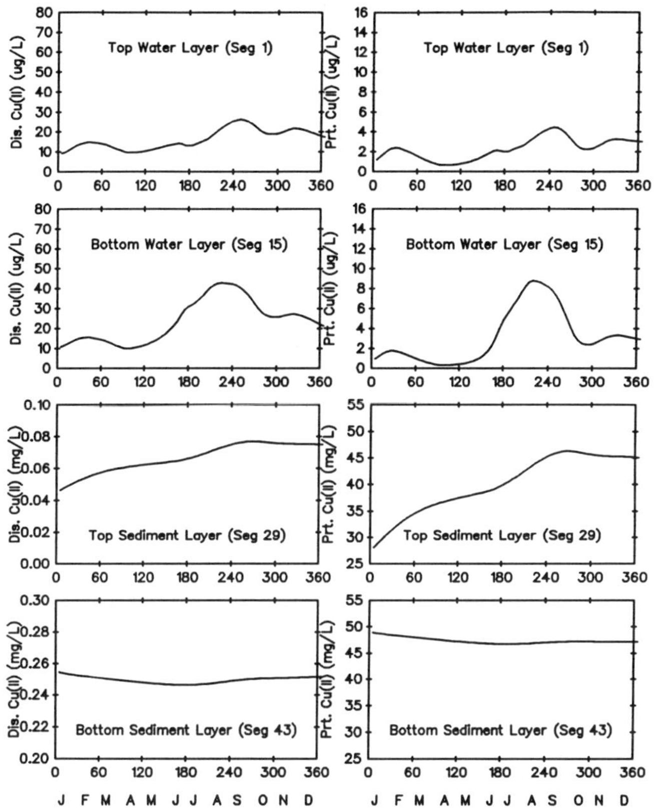

FIGURE 9. Copper (II) model results for the New York Bight.

months during summer and along the New Jersey shore.[65] Along with this strong southwesterly alongshore flow there exists a weaker cross-shelf circulation similar to estuarine flow. The lighter surface water tends to flow seaward above the pycnocline and more saline denser water flows toward the shore at the bottom. This estuarine flow results in a plume which generally makes a clockwise turn as it flows from the harbor and displaces further offshore during summer and the near bottom flow appears to be toward the shore. The fall plume stays closer to the mouth because of the southwesterly wind. It moves further offshore in winter and in spring remains in between that of summer and winter.[60] These seasonal flows through each segment are calculated from the circulation diagrams provided in the literature.[60,66]

The chemical data regarding the initial concentrations of dissolved and particulate Cu(II) and suspended solids in all the segments in the water column and the aerobic sediment layer were collected from report data.[67,68] In the anaerobic sediment layer, scarcity of data resulted in estimation of concentrations by assuming a linear spatial variation of the available data. The $CO_3^=$ and $SO_4^=$ data in the water column and sediment system were collected from Riley and Chester.[69] As no temporal data were found for these two components, the concentrations were assumed to be constant throughout the simulation. Data on pollutant loads entering the segments at the mouth of the Hudson-Raritan Bay were gathered from NOAA.[70] According to studies on the New York Bight sediments the settling of suspended materials takes place faster in the apex zone than the rest of the area.[68] Thus, during simulation, the apex zone was assumed to have faster settling rate for the suspended solids in the water column. Segar et al.[68] provided some

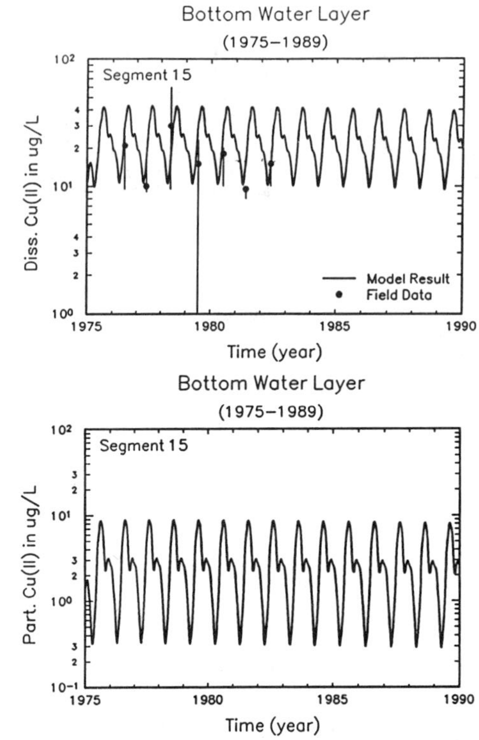

FIGURE 10. Long-term simulation results for segment no. 15 (bottom water layer).

information about the sediment burial rate in the New York Bight. The bioturbation rate was computed by using the labile particulate organic carbon in the aerobic layer and the dissolved oxygen concentration in the overlying water. These data were used in a one-year model run.

Figure 9 shows the dissolved and particulate concentrations of Cu(II) in the water column and sediment segments at the mouth of the Hudson-Raritan Bay. A seasonally variable input loadings were considered from the transect zone into these segments. This was well reflected by the model which shows a seasonal rise and fall of the metal concentrations in the water columns. Initially, the bottom sediment layer was assumed to be more contaminated with dissolved and particulate Cu(II) than the top sediment layer. Thus, the increase in top sediment layer particulate concentration is a result of bioturbation mixing causing more contaminated particles to be brought into the top sediment layer from the bottom sediment layer. The dissolved Cu(II) concentration in the same layer increases because of Cu(II) diffusing into this layer from the bottom sediment layer and repartitioning. The bottom sediment layer particulate Cu(II) concentration decreases due to burial.

The sediment column reacts very slowly to changes in the physical phenomenon involved in controlling the contaminant transport. Long-term simulations for a period from 1975 to 1990 were conducted. The Cu(II) inputs to the New York Bight were assumed to remain constant throughout this period. Figure 10 shows the model results for dissolved and sorbed Cu(II) in one of bottom water layer segments, compared with the limited data. Calculated dissolved Cu(II) concentrations match the observed. The seasonal rise and fall of dissolved Cu(II) concentrations evident from the data was mimicked by the model results. A similar trend is simulated by the model for the sorbed Cu(II) in the same segment. The aerobic sediment layer, being very thin, reaches a steady-state condition by rapidly approaching equilibrium (Figure 11). The long-term model simulations were performed by keeping the kinetic coefficients such as settling velocity

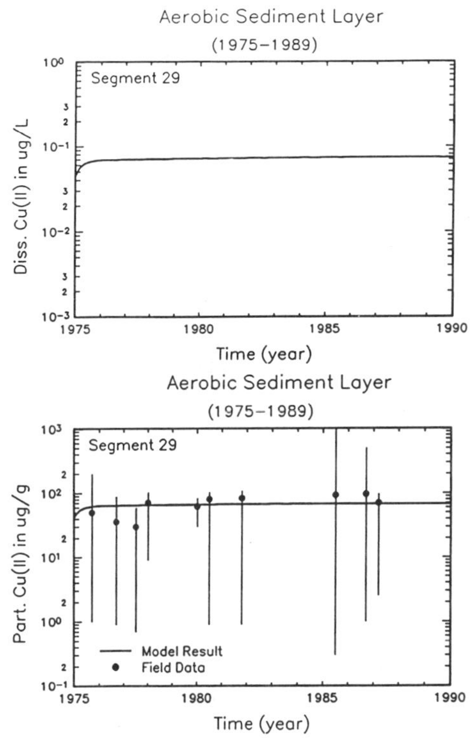

FIGURE 11. Long-term simulation results for segment no. 29 (aerobic sediment layer).

and dispersion coefficient in the water column, sediment burial rate, etc. constant for the simulation period. Thus, the subsequent match between the model results and field data offers additional substantiation of the model credibility.

Figure 12 shows the mean seasonal variation of dissolved Cu(II) species in the aerobic sediment at the mouth of the Hudson-Raritan Bay (segment 29). In the winter time, the predominant species is aqueous $Cu\ (CO_3)$ and free Cu^{2+}. In addition, aqueous $Cu\ SO_4$ and $Cu(CO_3)_2^{=}$ were present in smaller percentages. However, as the dissolved Cu(II) concentration increased with time, chemical composition of the interstitial water in the aerobic sediment showed a decrease in the aqueous $Cu\ (CO_3)$ and a simultaneous increase in free Cu^{2+} ion concentration until the dissolved Cu(II) concentration became more steady. In the anaerobic sediment (segment 43), the total Cu(II) concentration is very high because of accumulation over a prolonged period and showed a steadier dissolved and sorbed Cu(II) distribution with a slight decreasing trend almost the entire year except for a few days in the fall when it showed a slight increase (Figure 12). Thus, the distribution of Cu(II) species followed the similar trend showing a slight reduction of the free Cu^{2+} ions and increase in aqueous $Cu\ CO_3$ during the entire year except in the fall which registered some increase in free Cu^{2+} ion with the increase in dissolved Cu(II) concentration (Figure 12).

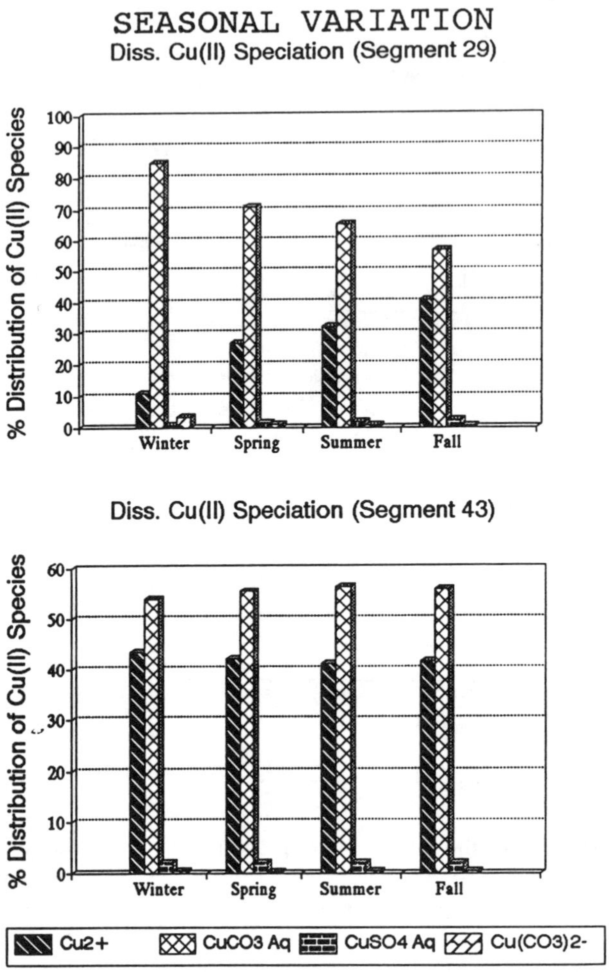

FIGURE 12. Seasonal variation of dissolved and sorbe Cu(II) species in aerobic and anaerobic sediment layers at the mouth of the Hudson-Raritan Bay.

REFERENCES

1. **Thomann, R. V. and Mueller, J. A.,** *Principles of Surface Water Quality Modeling and Control*, Harper & Row, New York, 1987, 8.
2. **Schnoor, J. L., Sato, C., McKechnie, D., and Sahoo, D.,** Processes, Coefficients, and Models for Simulating Toxic Organics and Heavy Metals in Surface Waters, U.S. Environmental Protection Agency, EPA/600/3-87/015, Athens, GA, 1987.
3. **U.S. EPA,** Technical Guidance Manual for Performing Waste Load Allocations, Book III: Estuaries, Part 1: Estuaries and Waste Load Allocation Models, 1990.
4. **Stumm, W. and Morgan, J. J.,** *Aquatic Chemistry*, 2nd ed., Wiley-Interscience, New York, NY, 1981.
5. **Lasaga, A. C.,** Rate laws of chemical reactions, in *Kinetics of Geochemical Processes; Reviews in Mineralogy*, Vol. 8, Lasaga, A.C. and Kirkpatrick, R.J., Eds., Mineralogical Society of America, 1981.
6. **Reinbold, K. A., Hassett, J. J., Means, J. C., and BanWart, W. L.,** Adsorption of Energy-Related Organic Pollutants: a Literature Review, U.S. Environmental Protection Agency, EPA- 600/3-79-086, 1979.

7. **Wu, S. C. and Gschwend, P. M.,** Sorption kinetics of hydrophobic organic compounds to natural sediments and soils, *Environ. Sci. Tech.*, 20, 717, 1986.
8. **Karickhoff, S. W.,** Sorption kinetics of hydrophobic pollutants in natural sediments, in *Contaminants and Sediments, Vol. 2: Fate and Transport, Case Histories, Modeling, Toxicity*, Baker, R.A., Ed., Ann Arbor Press, Ann Arbor, MI, 1980.
9. **Selim, H. M. and Amacher, M. C.,** A second-order kinetic approach for modeling solute retention and transport in soils, *Water Resour. Res.*, 24, 2061, 1988.
10. **McCarthy, J. F.,** Bioavailability and toxicity of metals and hydrophobic organic contaminants, in *Aquatic Humic Substances: Influence on Fate and Treatment of Pollutants*, Suffet, I.H. and MacCarthy, P., Eds., Advances in Chem. Series 219, American Chemical Society, Washington, D.C., 263, 1989.
11. **Chiou, C. T., Peters, L. J., and Freed, V. H.,** A physical concept of soil-water equilibrium for nonionic organic compounds, *Science*, 206, 831, 1979.
12. **Karickhoff, S. W., Brown, D. S., and Scott, T. A.,** Sorption of hydrophobic pollutants on natural sediments, *Water Resour.*, 13, 241, 1979.
13. **McCarthy, J. F. and Jimenez, B. D.,** Interactions between polycyclic aromatic hydrocarbons and dissolved humic material: binding and dissociation, *Environ. Sci. Tech.*, 19, 1072, 1985.
14. **Waldichuk, M.,** Exchange of pollutants and other substances between the atmosphere and the oceans, in *The Handbook of Environmental Chemistry, Vol. 1, Part D: Reactions and Processes*, Hutzinger, O., Ed., Springer-Verlag, New York, 1988, 113.
15. **Gucinski, H., Hardy, J. T., and Preston, H. R.,** Implications of toxic materials accumulating in the surface microlayer in Chesapeake Bay, in *Understanding the Estuary: Advances in Chesapeake Bay Research*, Lynch, M.P. and Krome, E.C., Eds., Chesapeake Bay Research Consortium Publ. No. 129,
16. **Eisenreich, S. J., Hollod, G. J., and Johnson, T. C.,** Atmosphere concentrations and deposition of polychlorinated biphenyls to Lake Superior, in *Atmosphere Pollutants in Natural Waters*, Eisenreich, S.J., Ed., Ann Arbor Press, Ann Arbor, MI, 1981, 425.
17. **Mackey, D., Shiu, W. Y., Bobra, A., Billington, J., Chau, E., Yeun, A., Ng, C., and Szeto, F.,** Volatilization of Organic Pollutants from Water, EPA-600/3-82-019, 1982.
18. **Liss, P. S. and Slatter, P. G.,** "Flux of gases across the air-sea interface," *Nature*, 247, 181, 1974.
19. **Thomas, R. G.,** Volatilization from water, in *Handbook of Chemical Property Estimation Methods*, Lyman, W.J., Reehl, W.F., and Rosenblatt, D.H., Eds., McGraw-Hill, New York, 1982, 15.
20. **Adamson, A. W.,** *Physical Chemistry of Surfaces*, 4th ed., John Wiley & Sons, New York, 1982.
21. **Rosen, M. J.,** *Surfactants and Interfacial Phenomena*, 2nd ed., John Wiley & Sons, New York, 1989.
22. **Gibson, V. R.,** Vertical distribution of estuarine phytoplankton, in the surface microlayer and at 1 meter and fluctuations in abundance caused by surface adsorption of monomolecular films, unpublished M.S. thesis, University of Virginia, Charlottesville, VA, 1977.
23. **Hardy, J. T., Crecelius, B. A., Antrim, L. S., Kiesser, S. L., Broadhurst, V. L., Boehm, P. D., and Steinhauer, W. G.,** Aquatic Surface Microlayer Contamination in Chesapeake Bay, report prepared by Battelle, Pacific Northwest Laboratory, Richland, WA for Maryland Power Plant Research Program (PPRP-100), 1987.
24. **VIMS,** 1988 Annual Report for Period Ending June 20, 1988, 41st report of VIMS and School of Marine Sciences of the College of William & Mary, Glouchester, VA, 1988.
25. **Sanders, J. G. and Riedel, G. F.,** Chemical and physical processes influencing bioavailability of toxics in estuaries, in *Perspectives on the Chesapeake Bay: Advances in Estuarine Sciences*, Lynch, M.P. and Krome, E.C., Eds., Chesapeake Bay Consortium Publ. No. 127, Document No. CBP/TRS 16/87, 1987.
26. **McCarthy, J. F.,** Bioavailability and toxicity of metals and hydrophobic organic contaminants, in *Aquatic Humic Substances: Influence on Fate and Treatment of Pollutants*, Suffet, I.H. and MacCarthy, P., Eds., Advances in Chem. Series 219, American Chemical Society, Washington, D.C., 1989, 263.
27. **O'Donnel, J. R., Kaplan, B. M., and Allen, H.E.,** Bioavailability of trace metals in natural water, in *Aquatic Toxicology and Hazard Assessment: 7th Symposium*, Cardwell, R.D., Purdy, R., and Bahner, R.C., Eds., ASTM Special Publ. 854, Philadelphia, PA, 1987, 485.
28. **Harrison, F. L.,** Effect of the physicochemical form of trace metals on their accumulation by bivalve molluscs, in *Chemical Modeling in Aqueous Systems: Speciation, Sorption, Solubility, and Kinetics*, Jenne, E.A., Ed., American Chemical Society Symp. Series 93, ACS, Washington, D.C., 1979, 611.
29. **Louma, S. N., and Bryan, G. W.,** "Trace metal bioavailability: modeling chemical and biological interactions of sediment-bound zinc," in *Chemical Modeling in Aqueous Systems: Speciation, Sorption, Solubility, and Kinetics*, Jenne, E. A., Ed., American Chemical Society Symp. Series 93, Washington, D.C., 577, 1979.
30. **Ball, J. W., Nordstrom, D. K., and Zachmann, D. W.,** WATEQ4F — a personal computer FORTRAN translation of the geochemical model WATEQ with revised data base, U.S. Geological Survey Open File Report 87–50, 1987.

31. **Felmay, A. R., Girvin, D. C., and Jenne, E. A.,** MINTEW — a computer program for calculating aqueous geochemical equilibria, U.S. EPA, Athens, GA, EPA-600/3-84-032, 1984.
32. **Allison, J. D., Brown, D. S., and Novo-Gradac, K. J.,** MINTEQA2/PRODEFA2, a Geochemical Assessment Model for Environmental Systems: Version 3.0 User's Manual. U.S. EPA Environmental Research Laboratory, Athens, GA, 1990.
33. **Fontaine, T. D.,** A non-equilibrium approach to modeling toxic metal speciation in acid, aquatic systems, in *Modelling the Fate and Impact of Toxic Substances in the Environment*, Jorgensen, S.E., Ed., Devel. Environ. Model No. 6, Elsevier, New York, 1984, 85.
34. **Perdue, E. M.,** Effects of humic substances on metal speciation, in *Aquatic Humic Substances: Influence on Fate and Treatment of Pollutants*, Advances in Chem. Series 219, Suffet, I.H. and MacCarthy, P., Eds., American Chemical Society, Washington, D.C., 1989, 281.
35. **Clement Associates, Inc.,** Chemical, Physical, and Biological Properties of Compounds Present at Hazardous Waste Sites, Final report prepared for U.S. EPA, 1985.
36. **Atlas, R. M. and Bartha, R.,** *Microbial Ecology: Fundamentals and Applications*, 2nd ed., Benjamin/ Cummings, Menlo Park, CA, 1987.
37. **Helz, G. R. and Kieber, R. J.,** The role of photochemistry in the bioavailability of toxic substances in estuaries; implications for monitoring, in *Understanding the Estuary: Advances in Chesapeake Bay Research*, Lynch, M.P. and Krome, E.C., Eds., Ches. Research Consort. Publ. No. 129, 1988.
38. **Mill, T. and Mabey, W.,** Hydrolysis of organic chemicals, in *The Handbook of Environmental Chemistry, Vol. 1, Part D: Reactions and Processes*, Hutzinger, O., Ed., Springer-Verlag, New York, 1988, 71.
39. **Parkhurst, D. L., Thorsterson, D. C., and Plummer, L. N.,** PHREEQE — a computer program for geochemical calculation, U.S. Geological Survey Water Resources Investigations Report 80-96, 1980.
40. **Mehta, A. J., Hayter, E. J., Parker, W. R., Krone, R. B., and Teeter, A. M.,** Cohesive sediment transport. I: Process description, *J. Hydraul. Eng.*, 115, 1076, 1989.
41. **Mehta, A. J., McAnally, W. H., Jr., Hayter, E. J., Teeter, A. M., Schoelhamer, D., Heltzel, S. B., and Cary, W. P.,** Cohesive sediment transport. II: Applications, *J. Hydraul. Eng.*, 115, 1094, 1989.
42. **Mills, W. B., Porcella, D. B., Ungs, M. J., Gherini, S. A., Summers, K. V., Mok, L., Rupp, G. L., and Bowie G. L.,** Water Quality Assessment: A Screening Procedure for Toxic and Conventional Pollutants in Surface and Ground Water, Part 1, U.S. Environmental Protection Agency, Athens, GA, EPA/00/6-85/002a, 1985, 5.
43. **Lung, W.S., Rapaport, R. A., and Franco, A. C.,** Predicting concentrations of consumer product chemicals in estuaries, *Environ. Toxicol. Chem.*, 9, 1127, 1990.
44. **Lung, W. S.,** Advective acceleration and mass transport in estuaries, *ASCE J. Environ. Eng.*, 112, 874, 1986.
45. **Lung, W. S. and O'Connor, D. J.,** Two-dimensional mass transport in estuaries, *ASCE J. Hydraul. Eng.*, 110, 1340, 1984.
46. **O'Connor, D. J.,** Models of sorptive toxic substances in freshwater systems. I. basic equations, *ASCE J. Environ. Eng.*, 114, 507, 1988.
47. **Rapaport, R. A.,** Prediction of consumer product chemical concentrations as a function of publicly owned treatment works treatment type and riverine dilution, *Environ. Toxicol. Chem.*, 7, 107, 1988.
48. **De Henau, H., Matthijs, E., and Hopping, W. D.,** Linear alkylbenzene sulfonates (LAS) in sewage sludges, soils and sediments: Analytical determination and environmental considerations, *Int. J. Environ. Anal. Chem.*, 26, 279, 1986.
49. **Painter, H. A. and Zabel, T. F.,** *Review of the Environmental Safety of LAS*, WRC Medmenham, Marlow, England, 1988.
50. **Vives-Rego, J., Vaque, M. D., Leal, J. S., and Parra, J.,** Surfactants biodegradation in sea water, *Tenside Surfact. Det.*, 24, 210, 1987.
51. **Shimp, R. J.,** LAS biodegradation in estuaries, *Tenside Surfact. Det.*, in press.
52. **Selleck, R. E., Glenne, B., and Pearson, E. A.,** A comprehensive study of San Francisco Bay, Final Report, Vol VII: A model of mixing and dispersion in San Francisco Bay, SERL Report No. 67-1, 1966.
53. **Goodwin, C. R.,** Circulation of Tampa and Sarasota Bays, in *Tampa and Sarasota Bays: Issues, Resources, Status, and Management*, NOAA Estuary-of-the-Month Seminar Series No. 11, 1989, 49.
54. **Goodwin, C. R.,** Tidal-flow, circulation, and flushing changes caused by dredge and fill in Tampa Bay, Florida, U.S. Geological Survey Water Supply Paper 2282, Tampa, FL, 1987.
55. **Goodwin, C. R.,** Tidal-flow, circulation, and flushing changes caused by dredge and fill in Hillsborough Bay, FL, U.S. Geological Survey Water Supply Paper, 2376, 1991.
56. **O'Connor, D. J. and Farley, K. J.,** Preliminary Analysis of Kepone Distribution in the James River, Manhattan College, Environmental Engineering and Science Program, 1977.
57. **Ruppe, L. M. and Lung, W. S.,** Data and Transport Analysis of Chemicals in Partially Mixed Estuaries, University of Virginia, Department of Civil Engineering Environmental Engineering Research Report No. 2, 1990.
58. **O'Connor, D. J., Mueller, J. A., and Farley, K. J.,** Distribution of kepone in the James River Estuary, *ASCE J. Environ. Eng.*, 109, 396, 1983.

59. **Badruzzaman, A. B. M. and Lung, W. S.,** Assessing Cu(II) speciation and transport in the New York Bight, in *Estuarine and Coastal Modeling*, ASCE, New York, 1992, 466.
60. **HydroQual, Inc.,** Assessment of Pollutant Fate in New York Bight, 1989.
61. **Ambrose, R. B., Wool, T. A., Connolly, J. P., and Schnaz, R. W.,** WASP4, a Hydrodynamic and Water Quality Model — Model Theory, User's Manual, and Programmer's Guide, U.S. EPA, EPA/600/3-87/039, Athens, GA, 1987.
62. **Novo-Gradac, K. J. and Wang, P. F.,** Water Quality Modeling in San Francisco Bay, Part 1: Speciation Modeling of Copper, unpublished manuscript, 1991.
63. **Lung, W. S.,** Lake acidification model: practical tool, *ASCE J. Environ. Eng.*, 113, 900, 1987.
64. **Schnoor, J. L.,** Fate and transport of dieldrin in Coralville reservoir: residues in fish and water following a pesticide ban, *Science*, 211, 840, 1981.
65. **Hansen, D. V.,** Circulation, MESA New York Bight Atlas Monograph 3, New York Sea Grant Institute, Albany, NY, 1977.
66. **Stoddard, A.,** Mathematical Model of Oxygen Depletion in the New York Bight: An Analysis of Physical, Biological and Chemical Factors, in 1975 and 1976, Ph.D. dissertation, University of Washington, Seattle, WA, 1983.
67. **NMFS,** The Effects of Waste Disposal in the New York Bight, Section 5: Chemical Studies, 1976.
68. **Segar, D. A. and Cantillo, A. Y.,** Trace Metals in the New York Bight, Middle Atlantic Continental Shelf and New York Bight Special Symposia Proceedings, *Am. Soc. Limnol. Oceanogr.*, 2, 171, 1976.
69. **Riley, J. P. and Chester, R.,** *Introduction to Marine Chemistry*, Academic Press, New York, 1971.
70. **NOAA,** Contaminant Inputs to the New York Bight, Tech. Memo, ERL-MESA-6, U.S. Department of Commerce, 1972.

Chapter 6

MODELING SEDIMENT-WATER INTERACTIONS*

I. INTRODUCTION

Why should a water-quality model of an estuary include a sediment component? Because accurate, reliable models of estuaries cannot be constructed if sediments are ignored. Sediment oxygen demand (SOD) is a major contributor to the degradation of Chesapeake Bay.[1] Sediment releases of nutrients support the algal population of the Patuxent Estuary[2] and contribute to blooms in the tidal Potomac.[3] Calibration of a water-quality model of the tidal James River is impossible if sediment-water nutrient fluxes are ignored.[4]

Publishing an up-to-date monograph on sediment modeling is difficult. The pace of research in the area is much faster than the turnaround time of publications. Nevertheless, it is possible to present fundamental principles and expose the reader to developments that are most promising at this time. This chapter emphasizes practical "engineering" approaches to modeling sediment-water fluxes of ammonium (NH_4^+), nitrate (NO_3^-), phosphate (PO_4^{-3}), and dissolved oxygen (DO). These substances form the minimum set of constituents to be considered in coupling sediment-water interactions to estuarine eutrophication models.

II. CONCEPTUAL MODELS

A. DIAGENESIS AND OXYGEN EQUIVALENTS

Diagenetic processes start with the deposition of particulate organic matter (POM) at the sediment-water interface (Figure 1). POM is considered to be composed of carbohydrate (CH_2O), NH_4^+, and PO_4^{-3}. The term *diagenesis* applies to all processes that affect POM and its components subsequent to deposition.[5] For our purposes, the primary diagenetic process is the oxidation of organic carbon and concurrent release of NH_4^+ and PO_4^{-3} incorporated in POM. Oxidation of carbon requires simultaneous reduction of another substance. Numerous substances (referred to as electron acceptors[6]) may be reduced including oxygen, NO_3^-, ferric iron ($Fe(III)$), sulfate ($SO_4^=$) and carbon dioxide (CO_2).

Sequences of oxidation-reduction (redox) reactions, ordered according to the free energy released, may be found in standard texts. The sequence that occurs in sediments need not conform to theory due to redox reaction kinetics and the formation of microenvironments.[6] Moreover, accounting for each electron acceptor and the myriad of possible reactions is a practical impossibility. A crucial concept that bypasses the accounting of individual electron acceptors is the "oxygen equivalents" approach.[7] In this approach, the end products of oxidation of organic carbon are viewed as carbon dioxide and chemical oxygen demand (COD). Decomposition of POM is then expressed:

$$(CH_2O)_a(NH_4^+)_b(PO_4^{-3})_c \rightarrow dCO_2 + eCOD + bNH_4^+ + cPO_4^{-3} \tag{6.1}$$

where

a = moles CH_2O per mole organic matter
b = moles NH_4^+ per mole organic matter
c = moles PO_4^{-3} per mole organic matter
d = moles CO_2 produced per mole organic matter
e = moles O_2 required per mole organic matter oxidized

*Sections I to IV contributed by Carl F. Cerco.

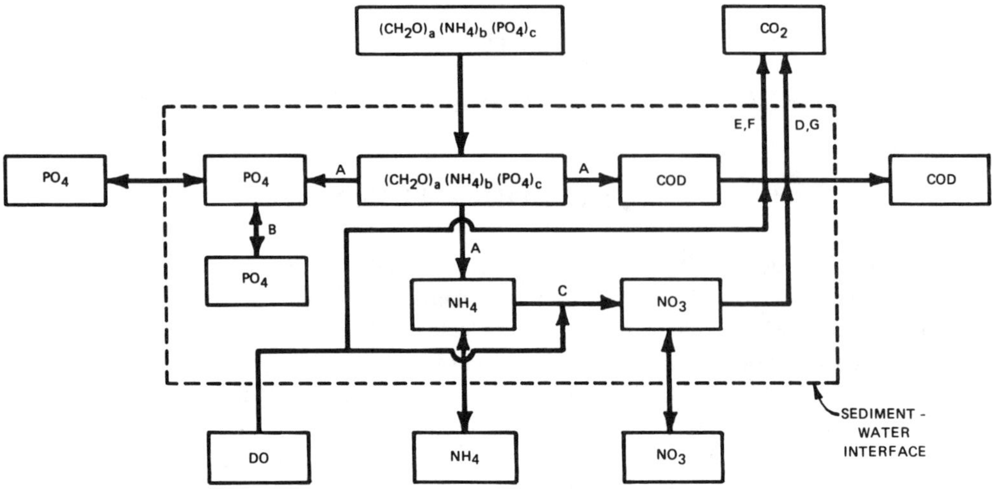

FIGURE 1. Conceptual sediment model.

The conceptual models considered here assume that all species of COD are transported at the same rate. Formation of solid- and gaseous-phase COD are ignored. More elegant treatments of these COD phases have been conducted.[7,8]

B. FATE OF COD

As viewed here, SOD is a two-step process involving the creation and exertion of an oxygen demand. In reality, SOD may be exerted in a single step through respiration:

$$O_2 + CH_2O \rightarrow CO_2 + H_2O \tag{6.2}$$

or in two steps, e.g., reduction of $SO_4^=$ and oxidation of HS

$$SO_4^= + 2\,CH_2O + H^+ \rightarrow HS^- + 2CO_2 + 2H_2O \tag{6.3}$$

$$2CO_2 + HS^- \rightarrow SO_4^= + H^+ \tag{6.4}$$

Respiration is referred to as biological sediment oxygen demand (BSOD) while oxidation of reduced intermediates is chemical sediment oxygen demand (CSOD). CSOD comprises 0 to 80% of total SOD.[9-12] The distinction is to some extent artificial since chemical oxidation of a reduced intermediate must be preceded by biological reduction of an electron acceptor.

COD may also be oxidized by reduction of NO_3^-. Simultaneous oxidation of organic carbon and reduction of NO_3^- is referred to as denitrification.

$$4NO_3^- + 4H^+ + 5CH_2O \rightarrow 2\,N_2 + 7H_2O + 5CO_2 \tag{6.5}$$

Abiotic reduction of NO_3^- to satisfy COD is also thermodynamically feasible.

$$5HS^- + 8NO_3^- + 3H^+ \rightarrow 5SO_4^= + 4N_2 + 4H_2O \tag{6.6}$$

NO_3^- reduction is unique because the gaseous end product, N_2, is not subject to oxidation. Hence NO_3^- reduction offers a pathway for oxidation of organic matter without the consumption of oxygen.

Oxidation of POM in sediments does not require oxygen. In the absence of oxygen, alternate electron acceptors may be employed. Production of methane, in which CO_2 is the electron acceptor, is always possible.

$$2CH_2O \rightarrow CH_4 + CO_2 \tag{6.7}$$

If the rate of oxygen transfer from the water column to the sediments is insufficient to oxidize the COD produced, export of COD to the water column occurs.

C. FATE OF AMMONIA

If sufficient oxygen is available, biological oxidation (nitrification) of sediment NH_4^+ occurs.

$$2O_2 + NH_4^+ \rightarrow NO_3^- + 2H^+ + H_2O \tag{6.8}$$

The Redfield ratios and redox stoichiometry indicate that nitrogenous demand comprises 20% of total SOD, provided both carbon and nitrogen are completely oxidized using oxygen as the electron acceptor. The fraction actually represented by nitrification may be more or less, however, since complete oxidation of NH_4^+ and carbon seldom take place. Detailed consideration of the proportion of total SOD represented by NSOD may be found in the literature.[8] Sediment uptake of NH_4^+ occurs occasionally.[13,14] The fate of NH_4^+ taken up by sediments is not clear, but NH_4^+ uptake is an order of magnitude less than release rates observed in the same sediments. NH_4^+ may also sorb to sediment particles.[15] The primary importance of sorption is to increase the fraction of NH_4^+ lost to burial in deep sediments.

D. FATE OF NITRATE

NO_3^- may be taken up or released by sediments. Direction of movement depends upon the concentration in the overlying water, extent of nitrification within the sediments, and the rate of POM decomposition. In the sediments, the primary fate of NO_3^- is reduction to gaseous nitrogen. Reduction of NO_3^- is a process of interest in estuarine eutrophication since production of N_2 gas comprises a loss of nitrogen from the system. This loss may contribute to the nitrogen limitation of primary production commonly observed in estuaries.[16] Absent NO_3^- reduction, the only pathway for removal of nitrogen is burial to the deep sediments.

E. FATE OF PHOSPHATE

PO_4^{-3} released from POM to sediment interstitial waters undergoes a sorption-desorption reaction with inorganic particles. In an anoxic environment, the mass of sorbed PO_4^{-3} vastly exceeds the mass in interstitial water.[17] Sediments with a large sorption capacity may strip PO_4^{-3} from the overlying water.[18] Sudden decrease in the sorption capacity of sediment particles results in the bursts of PO_4^{-3} release observed when water overlying sediments becomes anoxic.[19] A primary cause of anoxia-induced PO_4^{-3} release is bacterial reduction of *Fe(III)* to *Fe(II)*.[20] In the transformation of iron from a solid to dissolved phase, PO_4^{-3} sorbed to the solids is released to the interstitial waters and subsequently to the water column.[21] In tidal freshwater, PO_4^{-3} may also be stripped from particles at pH > 9[3,22] but the buffering of pH in seawater precludes this process from being important in the saline portions of estuaries.

III. EMPIRICAL MODELS

A. UTILITY OF EMPIRICAL MODELS

Once the processes that occur in sediments are understood, models of varying degrees of complexity can be formulated. The simplest models are empirical relationships that calculate sediment-water nutrient and oxygen fluxes as functions of conditions in the water column. Water column properties considered here are temperature, nutrient concentration, and DO. These factors are fundamental to the determination of sediment-water flux, are commonly measured, and are incorporated in eutrophication models. Empirical models that predict flux based on these independent variates are readily coupled with eutrophication models that predict temperature and concentrations of nutrients and DO.

B. FORMULATION OF EMPIRICAL MODELS

Sufficient laboratory and *in situ* studies are available upon which to base formulations of empirical models. In the case of NH_4^+, sediment release predominates over uptake.[13,14,19,23-27] Release increases as a function of temperature[13,14,27] and at low DO concentrations.[19] These observations are described by the model:

$$F = F_o q^{T-20} e^{\alpha DO} \tag{6.9}$$

where

F = sediment-water nutrient flux (mg/m²/d)
F_o = NH_4^+, PO_4^{-3} release rate at 20°C (mg/m²/d)
α = constant that expresses effect of DO on flux (m³/g)
θ = constant that expresses effect of temperature on sediment-water flux
T = temperature (°C)

In eutrophic systems characterized by substantial NO_3^- concentrations, flow of NO_3^- is predominantly from the water column to the sediments[14,18,28] and is proportional to water column concentration.[28-30] Sediment uptake of NO_3^- is enhanced by high temperature[29,31-33] and by low DO.[28] These observations are described by the model:

$$F = bC_w q^{T-20} e^{\alpha DO} \tag{6.10}$$

where

C_w = concentration of NO_3^-
PO_4^{-3} = in water column

Phosphorus moves in both directions across the sediment-water interface.[13,23-25] In some instances, sediments buffer PO_4^{-3} concentration in the overlying water.[34] When concentration exceeds an equilibrium concentration, movement of PO_4^{-3} is from water to sediment. When concentration is less than equilibrium, PO_4^{-3} moves from the sediments to the water column. The tendency for sediment release of PO_4^{-3} to increase as DO decreases is well known[35-37] as is the tendency for PO_4^{-3} release to increase as a function of temperature.[13,14,35-37] These observations are described by the model:

$$F = F_o q^{T-20} e^{\alpha DO} - bC_w \tag{6.11}$$

where

β = proportionality constant between flux and concentration (mm/d)

SOD increases as a function of temperature[13,14,38,39] and decreases as the supply of oxygen to the sediments is limited.[19,25,33,39] These observations are described by the model:

$$F = F_o q^{T-20} (1 - e^{\alpha DO}) \tag{6.12}$$

TABLE 1
Parameters in Empirical Models[24,40]

	NH_4^+	NO_3^-	PO_4^{-3}	SOD
F_o (mg m^{-2} d^{-1})	108		10.2	1.9[a]
θ	1.070	1.055	1.0	1.046
α(l/mg)	−0.066	−0.083	−0.184	−0.340
β(mm/d)		110	134	

[a] g^{-2} d^{-1}.

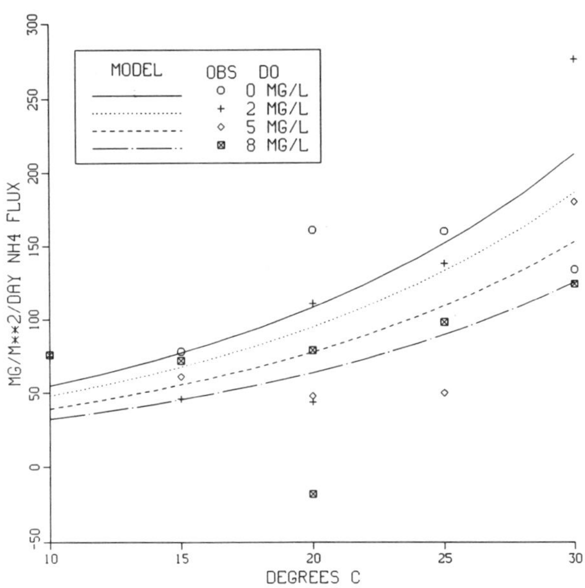

FIGURE 2. Observed and modeled sediment ammonium release.

The empirical models described by Equations 6.9 to 6.12 have been fitted to observations collected in Gunston Cove,[24,40] a tidal freshwater tributary of Chesapeake Bay. Parameter values are presented in Table 1 and model performance is compared to observations in Figures 2 to 5.

C. INTERPRETATION OF EMPIRICAL MODELS

Properly formulated empirical models are consistent with the conceptual model (Figure 1) and lend insight into the processes that occur in the sediments studied. Temperature enhancement of NH_4^+ release, NO_3^- uptake, and SOD indicates a temperature-induced increase in the breakdown of POM (Reaction A in Figure 1). Increased production of NH_4^+ results in export to the water column. Enhanced production of COD causes transport of DO and NO_3^- into the sediments and subsequent reduction of these substances to satisfy the COD. Both the conceptual model and the preponderance of observations indicate PO_4^{-3} release should increase as a function of temperature but observations in the study system show no effect. This result indicates the sorption capacity of the sediments (Reaction B in Figure 1) is sufficient to adsorb the PO_4^{-3} production associated with temperature-induced breakdown of POM. Hence, no dependence of release on temperature occurs.

Enhancement of NH_4^+ release at low DO is most likely due to suppression of nitrification (Reaction C in Figure 1). NH_4^+ that is no longer nitrified is instead released to the water column.

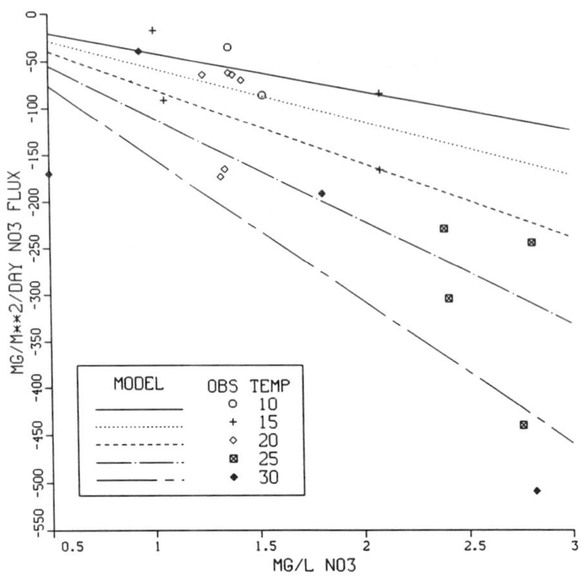

FIGURE 3. Observed and modeled sediment nitrite uptake.

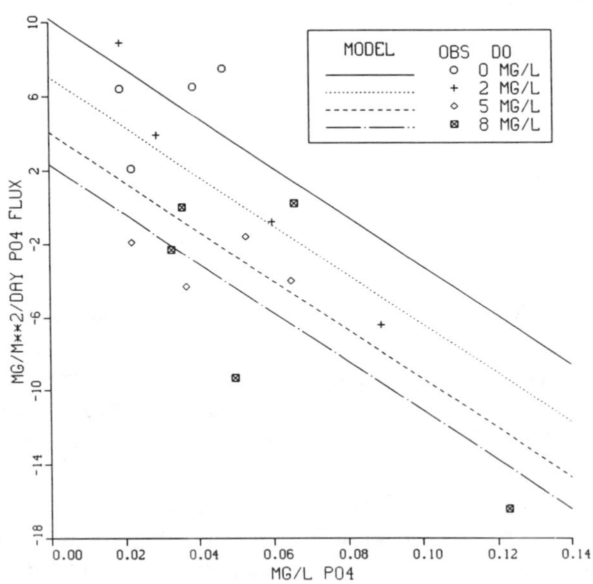

FIGURE 4. Observed and modeled sediment-water phosphate flux.

Increased NO_3^- uptake and diminished SOD at low DO indicate that employment of NO_3^- as an electron acceptor (Reactions D and G in Figure 1) increases as the supply of oxygen is limited.

Higher concentrations of NO_3^- in the water column result in an increased concentration gradient at the sediment-water interface and increased diffusion of NO_3^- into the sediments. Once in the sediments, the NO_3^- is consumed in denitrification or exertion of COD (Reactions D and G in Figure 1). Substitution of appropriate parameters into Equation 6.11 indicates that during the study period the sediments buffered PO_4^{-3} in DO-saturated overlying water at 0.018 g $PO_4 - P/\text{m}^3$. When concentration exceeded this level, sediments removed PO_4^{-3} from the

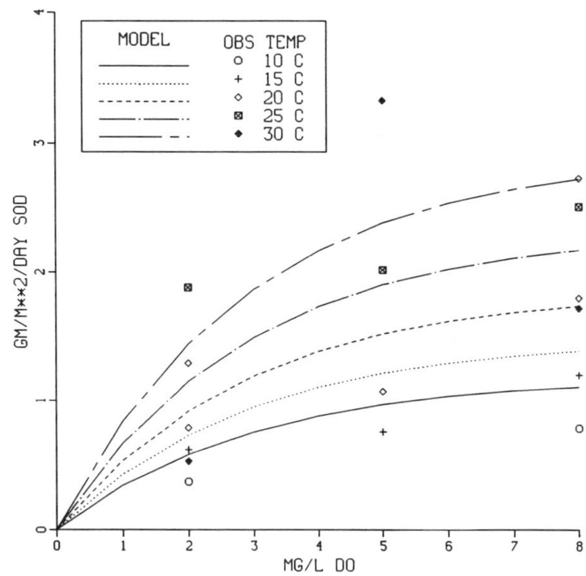

FIGURE 5. Observed and modeled sediment oxygen demand.

overlying water. Reference to Equation 6.11 or Figure 4 indicates that the equilibrium concentration was a function of DO in the overlying water. As DO decreased, the concentration at which zero flux occurred increased to 0.07 g $PO_4 - P$/m^3 under anoxic conditions.

D. PRECAUTIONS FOR EMPIRICAL MODELS

Empirical models can be of great utility in the calibration and application of eutrophication models.[22] Reliable employment of empirical models requires site-specific parameter evaluation, however. Utilization of reported parameters, without verification, can lead to spurious results. For example, parameters from other systems indicate a nonexistent relationship of temperature and phosphorus release in Gunston Cove. A second danger lies in omitting a significant independent variate from an empirical model. Subsequent to the reported study, sediment PO_4^{-3} release was found to be closely correlated with pH,[3,22] a factor not originally considered.

Parameter evaluation and omission of significant variates are problems with any model, empirical or mechanistic. The primary shortcoming of empirical models is they do not answer a fundamental question of interest in eutrophication studies: "What is the effect on sediment-water fluxes of altering POM input to the sediment?" To address this issue, more complex diagenetic models are required.

IV. DIAGENETIC MODELS

A. BASIC DIAGENETIC EQUATION

Complete diagenetic models account for both transport and transformations within the sediments. Diagenetic models have long been employed in marine biogeochemistry[5,41-45] but have found limited application in engineering. Lack of engineering application is due to the complexity of diagenetic models, the quantity of data required to apply the models, and the focus of the models. Diagenetic models emphasize the spatial distribution of substances within the sediments. Engineering applications focus on flux across the sediment-water interface. Nevertheless, diagenetic models are required to address the issue of alterations in the rate of POM input to the sediments. Applications of diagenetic models are finding their way into practice[46] and their employment is certain to increase.

The basic diagenetic equation is a one-dimensional advection-diffusion equation for a non-conservative substance:

$$\frac{\partial C}{\partial t} = \frac{\partial}{\partial y}\left(D_b \frac{\partial C}{\partial y}\right) + \frac{D_m}{1+K}\frac{\partial^2 C}{\partial y^2} - W\frac{\partial C}{\partial y} + \frac{\Sigma S}{1+K} \tag{6.13}$$

where

D_b = bioturbation diffusion coefficient (L^2/T)
K = dimensionless adsorption coefficient
D_m = molecular diffusion coefficient (L^2/T)
W = accretion rate at sediment-water interface (L/T)
S = internal source of dissolved substance C ($M\ L^{-3}\ T^{-1}$)

Equation 6.13 is appropriate for sediments in which porosity and adsorption are spatially invariant, in which no externally impressed water flow exists, and in which no sources or sinks of adsorbed C are active. Derivation and more elaborate forms of Equation 6.13 can be found in another document.[5]

B. DIAGENETIC NUTRIENTS AND OXYGEN MODEL

Diagenetic equations for six constituents are required to model the fluxes of nutrients and oxygen with which we are concerned. A simplification that facilitates both solution and interpretation is the assumption of steady-state. Under steady-state conditions, the equations become:

1. Particulate Organic Carbon

$$W\frac{dPOC}{dy} = \frac{d}{dy}\left(D_b \frac{dPOC}{dy}\right) - K_{poc} POC \tag{6.14}$$

where

K_{poc} = POC decomposition rate (T^{-1})

Implicit in the formulation of Equation 6.14 is the concept that sediment particles are subject to mixing by bioturbation but not diffusion.

2. Chemical Oxygen Demand

$$\begin{aligned} W\frac{dCOD}{dy} = \frac{d}{dy}\left(D_b \frac{dCOD}{dy}\right) + D_m \frac{d^2 COD}{dy^2} + A_{cp} K_{poc} \frac{1-\phi}{\phi} POC \\ -K_{cod}\left[\frac{DO}{K_{hdc}+DO} + \frac{NO_3^-}{K_{hnc}+NO_3^-}\right] COD \end{aligned} \tag{6.15}$$

where

A_{cp} = mass COD produced per mass POC decomposed
K_{cod} = COD oxidation rate (T^{-1})
K_{hdc} = half-saturation concentration of DO required for COD oxidation ($M\ L^{-3}$)
K_{hnc} = half-saturation concentration NO_3^- required for COD oxidation ($M\ L^{-3}$)

Dissolved substances, including COD, are subject to mixing by both bioturbation and molecular diffusion. The term containing ϕ accounts for the conversion from mass per unit particle volume to mass per unit porewater volume.

3. Dissolved Oxygen

$$W\frac{dDO}{dy} = \frac{d}{dy}\left(D_b\frac{dDO}{dy}\right) + D_m\frac{d^2DO}{dy^2} - K_{cod}\frac{DO}{K_{hdc}+DO}COD - K_{nh4}A_{dn}\frac{DO}{K_{hdn}+DO}NH_4^+ \tag{6.14}$$

where

K_{nh4} = nitrification rate (T^{-1})
A_{dn} = mass DO consumed per mass NH_4^+ nitrified
K_{hdn} = half-saturation concentration of D required for nitrification ($M\ L^{-3}$)

4. Ammonia

$$W\frac{dNH_4^+}{dy} = \frac{d}{dy}\left(D_b\frac{dNH_4^+}{dy}\right) + \frac{D_m}{1+K'}\frac{d^2NH_4^+}{dy^2} + A_{np}\frac{K_{poc}}{1+K'}\frac{1-\phi}{\phi}POC - \frac{K_{nh4}}{1+K'}\frac{DO}{K_{hdn}DO}NH_4^+ \tag{6.17}$$

where

A_{np} = nitrogen-to-carbon mass ratio of POM
K' = dimensionless adsorption coefficient of NH_4^+ to sediment particles

In this equation the adsorption coefficient, K', modifies the molecular diffusion and source-sink terms. A qualitative description of the effect of adsorption on sources and sinks is that a fraction of NH_4^+ released to porewaters (e.g., through POM decomposition) is lost due to adsorption by solids. NH_4^+ removed from porewaters (e.g., through nitrification) is partially replaced by desorption from solids.

5. Nitrate

$$W\frac{dNO_3^-}{dy} = \frac{d}{dy}\left(D_b\frac{dNO_3^-}{dy}\right) + D_m\frac{d^2NO_3^-}{dy^2} - A_{nc}K_{cod}\frac{NO_3^-}{K_{hdn}+NO_3^-}COD + K_{nh4}\frac{DO}{K_{hdn}+DO}NH_4^+ \tag{6.18}$$

where

A_{nc} = mass nitrate nitrogen reduced per mass COD oxidized

6. Phosphate

$$W\frac{d(1+K'')PO_4^{-3}}{dy} = \frac{d}{dy}\left(D_b\frac{d(1+K'')PO_4^{-3}}{dy}\right) + D_m\frac{d^2PO_4^{-3}}{dy^2} + A_{pp}K_{poc}\frac{1-\phi}{\phi}POC \tag{6.19}$$

where K'' = dimensionless adsorption coefficient of PO_4^{-3} to sediment particles, and A_{pp} = phosphorus-to-carbon ratio of POM.

Treatment of PO_4^{-3} sorption differs from NH_4^+ since the PO_4^{-3} sorption coefficient is not spatially constant. A larger fraction of PO_4^{-3} sorbs to oxic sediments than anoxic sediments. The oxygen dependence of PO_4^{-3} sorption is given by:

$$K = K_{anox} + K_{ox}\left(1 - e^{-K_{sd}DO}\right) \tag{6.20}$$

where

K_{anox} = dimensionless adsorption coefficient of PO_4^{-3} in anoxic sediments
K_{ox} = dimensionless adsorption coefficient of PO_4^{-3} in oxic sediments
K_{sd} = constant that expresses the effect of DO on PO_4^{-3} adsorption (L^3/M)

C. BOUNDARY CONDITIONS AND SOLUTION SCHEME

The POC boundary condition at the sediment-water interface is that flux across the interface is equal to the deposition rate. The boundary condition for the remaining substances is specified as the concentration at the sediment-water interface:

$$C_{y=0} = C_{yo} \tag{6.21}$$

For all substances, a second boundary condition is specified at the bottom of the active sediment layer:

$$\frac{dC}{dy} = 0 \quad \text{at} \quad y = y_b \tag{6.22}$$

Fluxes of nutrients, COD, and oxygen across the sediment-water interface are computed by a mass balance. At steady state, flux equals to the sum of all internal sources, less internal sinks, less burial to the deep sediments:

$$J_c = \int_0^{y_b} \phi S_{oc}\, dy - \int_0^{y_b} \phi S_{ic}\, dy - WC_{yb}(1 + K) \tag{6.23}$$

where

J_c = flux of substance across sediment-water interface ($M\,L^{-2}\,T^{-1}$)
S_{oc} = internal source of dissolved substance C ($M\,L^{-3}\,T^{-1}$)
S_{ic} = internal sink of dissolved substance C ($M\,L^{-3}\,T^{-1}$)
C_{yb} = concentration in interstitial water at bottom of active sediment layer ($M\,L^{-3}$)

An alternative method of computing flux is Fick's law of diffusion but no definitive method exists to quantify the concentration gradient at the sediment-water interface. The mass balance method eliminates the need to compute the gradient.

Analytical solutions to specific applications of diagenetic equations exist.[7,8,41-45] The number of constituents and the desirability of a general solution here, however, lead to solution by a finite-difference analog to the differential equations.

D. PARAMETER VALUES

Literature and model parameter values are presented in Table 2. Two values of K_{poc} are employed in the model: a larger value in the bioturbated zone (0 to 2.5 cm) and a lesser value below. This approach reflects the concept that organic matter close to the sediment-water interface is more recently deposited and more labile than matter buried below. An alternative to this approach is the division of organic matter into several components, each having a different decay rate.[5]

Definition of an active sediment layer is arbitrary. For this model, the active zone is extended down until computed sediment-water fluxes show no reaction to incremental changes in the location of the lower boundary.

E. MODEL RESULTS

The initial model run is made with parameters as specified in Table 2. Boundary conditions of no nutrients, no COD, and DO saturation in the water column are specified. Predicted nutrient fluxes and SOD (Table 3) are within ranges observed in several estuaries (Table 4).

1. Dissolved Oxygen and COD

Approximately 90% of the oxygen demand represented by organic carbon deposition is exerted as SOD or returned to the water column as COD. The remainder is buried as POM or COD. The ratio of COD exported to SOD exerted is higher than reported values (0.32 to 1.0)[19,54,55] and could be diminished by including methane bubble formation in the model.[8] As DO in the water column declines, SOD also declines but COD export to the water column increases (Figure 6). This phenomenon has long been known[9] and recently reiterated.[7] The inverse relationship of SOD exerted and COD exported has substantial implications in eutrophication modeling. Under anoxic conditions, the oxygen demand of sediments is not eliminated but exported to the water column. Examination of the effect of organic loading on SOD indicates a nonlinear response (Figure 7). A 50% reduction in loading results in approximately 25% reduction in SOD (Table 3). This phenomenon has been previously noted in an analytical model.[8] The difference between reduction in loading and corresponding reduction in SOD is reflected in a decline in COD export to the water column (Figure 7). The nonlinear response of SOD to loading reduction is a phenomenon that must be reflected in any management plan aimed at reducing SOD.

2. Ammonia and Nitrate

The fraction of nitrogen input to the sediments that is recycled to the water column depends on the oxygen concentration. Under anoxic conditions, 94% of the nitrogen input is recycled as NH_4^+ (Table 3). The balance is buried in deep sediments. Under DO-saturated conditions, 72% is returned to the water column, primarily as NH_4^+ but with some NO_3^- release as well. Release of NO_3^- indicates the occurrence of nitrification in the sediments but the NO_3^- released is less than the NH_4^+ nitrified. The NH_4^+ that is nitrified but not released as NO_3^- is denitrified to nitrogen gas. In the presence of DO saturation, 22% of the nitrogen input to the sediments is lost to denitrification, a result in agreement with reported values for coastal marine sediments.[56] The nitrogen fraction lost to denitrification also depends on the nitrogen loading to the sediments. Doubling the organic load diminishes the fraction lost to 14% of the nitrogen input (Table 3).

The dependence of nitrification and denitrification on DO leads to the concept of estuarine eutrophication as a vicious cycle. A decline in DO causes enhanced sediment NH_4^+ release. The NH_4^+ stimulates water-column nitrification and primary production. A fraction of the primary production settles to the sediments and manifests as SOD. The cycle completes itself when nitrogenous oxygen demand in the water and SOD further diminish DO and enhance sediment NH_4^+ release.

The diagenetic model conforms (Table 3) to the expectation that sediment uptake of NO_3^- increases in proportion to concentration in the water column.[28-30] Substantial NO_3^- release does not occur, however. In order for NO_3^- release to occur, the rate of production of NO_3^- by nitrification must exceed the rate of reduction to satisfy COD. This condition may exist ephemerally or in microenvironments but does not prevail in a horizontally-integrated, steady-state model.

TABLE 2
Parameters for Diagenetic Model

Parameter	Value	Unit	Reference	Comments	Model value
A_{cp}				Eq. 6.2	2.67
A_{dn}				Eq. 6.8	4.57
A_{nc}				Eq. 6.5	0.35
A_{np}				Redfield ratios	0.18
A_{pp}				Redfield ratios	0.024
D_b	8.6	$cm^2\ d^{-1}$	41	Upper 3.5 cm, North Sea	10
D_m	0.5	$cm^2\ d^{-1}$	5	Fine-grain marine sediments	0.5
J_{poc}	0.5–1.6	$g\ m^{-2}\ d^{-1}$	47	Upper Chesapeake Bay	1.0
K_{anox}	1.3–1.9	dimensionless	17	Anoxic marine sediments	1.5
K_{cod}	40	d^{-1}	48	1st-order oxidation rate	20
	3.6–26		49	Sulfide in seawater	
K_{hdc}		$g\ m^{-3}$			0.2
K_{kdn}	0.3–1.0	$g\ m^{-3}$	50		0.2
K_{hnc}	0.22–7.0	$g\ m^{-3}$	50		0.2
K_{ox}	34	dimensionless	17	Oxic marine sediments	33.5
K_{nh4}	0.06–0.78	d^{-1}	51	River water, sewage effluent	30
K_{poc}	0.0005–0.001	d^{-1}	52	Danish marine sediments	0.001, 0.0001
K_{sd}		$m^3\ g^{-1}$			2.3
K'	1.45	dimensionless	15	Marine sediments	1.5
W	0.45–0.60	$cm\ yr^{-1}$	53	Chesapeake Bay	0.5
y_b					50
ϕ	0.68–0.93	dimensionless	15	Tidal Potomac River	0.8

TABLE 3
Diagenetic Model Nutrient and COD Fluxes, SOD

NH_4^+ ($mg\ m^{-2}\ d^{-1}$)	NO_3^- ($mg\ m^{-2}\ d^{-1}$)	PO_4^{-3} ($mg\ m^{-2}\ d^{-1}$)	COD ($g\ m^{-2}\ d^{-1}$)	SOD ($g\ m^{-2}\ d^{-1}$)	Comments
123	7	22.6	1.49	1.13	DO-saturated water column
169	0	22.6	2.55	0	Anoxic water column
280	8	45.1	3.67	1.50	$J_{poc} = 2\ g\ m^{-2}\ d^{-1}$ in DO-saturated water column
339	0	45.1	5.10	0	$J_{poc} = 2\ g\ m^{-2}\ d^{-1}$ in Anoxic water column
123	–15[a]	22.6	1.44	1.12	0.1 $g\ m^{-3}\ NO_3^-$-N in water column
117	–191[a]	22.6	1.04	1.03	1.6 $g\ m^{-3}\ NO_3^-$-N in water column

[a] Into sediment.

TABLE 4
Estuarine Sediment Nutrient Fluxes and SOD[a]

NH_4^+ ($mg\ m^{-2}\ d^{-1}$)	NO_3^- ($mg\ m^{-2}\ d^{-1}$)	PO_4^{-3} ($mg\ m^{-2}\ d^{-1}$)	SOD ($g\ m^{-2}\ d^{-1}$)	Estuarine system	Refs.
10 ~ 280	–40 ~ 100	–3 ~ 30	1.5 ~ 3.5	Chesapeake Bay	23
–1 ~ 90	–20 ~ 15	–7 ~ 31	0.1 ~ 2.6	Narragansett Bay	13
0 ~ 150	0 ~ 2	–6 ~ 34	0.6 ~ 2.4	Neuse and South Rivers, NC	25
–40 ~ 360	–100 ~ 80	–19 ~ 124	0.1 ~ 2.7	Potomac Estuary	26
–35 ~ 530	–230 ~ 30	1 ~ 220	0.5 ~ 4.1	Patuxent Estuary	14

[a] Nutrient flux < 0 (into sediment).

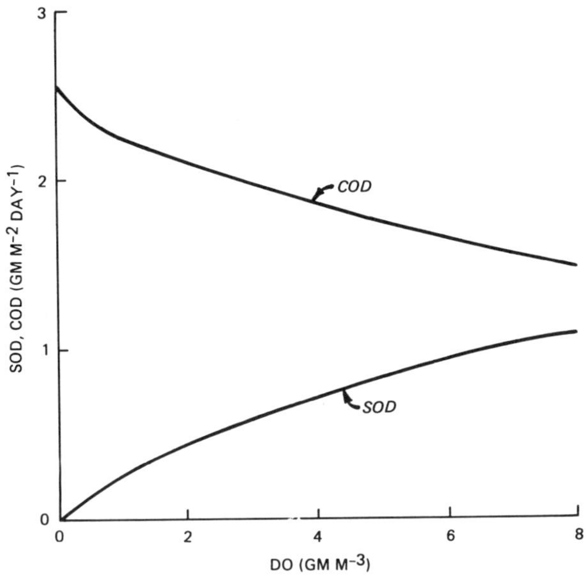

FIGURE 6. Effect of dissolved oxygen concentration on SOD and sediment COD export.

3. Phosphate

No difference in modeled oxic vs. anoxic PO_4^{-3} release is evident (Table 3). This behavior is readily understood. At steady state, the flux of particulate organic phosphorus to the sediments must be balanced by return to the water column, less any sinks. The only sink of PO_4^{-3} is burial to the anoxic, deep sediments. The burial rate is independent of oxygen concentration in the water column, however. Hence, at steady state, sediment phosphorus release is dependent upon the deposition and burial rates, not the oxygen concentration. The commonly-observed pulse phosphorus release that accompanies the onset of anoxia must be largely comprised of phosphorus previously mineralized and sorbed to particles in the oxic portions of sediments. Phosphorus release in excess of particulate input cannot be sustained.

V. CASE STUDY: SEDIMENT-WATER INTERACTIONS IN THE PATUXENT ESTUARY

Following the description of the sediment processes, a water quality model with sediment-water interactions incorporated for the Patuxent Estuary[57] is presented to demonstrate a successful application of the approach described. One of the difficulties in sediment modeling is the availability of data. Although some sediment oxygen demand and nutrient flux data in the Patuxent Estuary were available, no interstitial water concentrations were measured. Further, sediment porosity and compaction information was lacking. Such data limitation does not warrant a comprehensive sediment model to be interfaced with the water column model. In addition, it was desirable to keep the computation to manageable proportions. Therefore, the attempt to model the sediment system was limited to the quantification of sediment-water fluxes. Essentially, a diagenesis model was developed for the Patuxent Estuary.[57]

The settling velocity of particulates across the sediment-water interface is the net velocity on a tidally-averaged basis. A more comprehensive treatment of the particle settling into the sediment would include not only a downward velocity but an upward resuspension velocity as well. In the Patuxent model, the single velocity can be thought of as the settling velocity which represents the new flux to the sediment due to the difference between the downward settling flux and the upward resuspension flux.

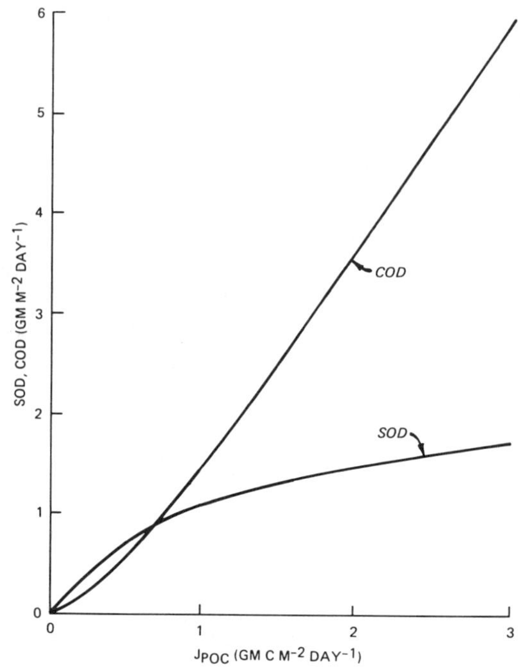

FIGURE 7. Effect of particulate organic carbon flux on SOD and sediment COD export.

The thickness of the sediment layer for the Patuxent Estuary model is another important parameter. Such a thickness should adequately reflect the thickness of the active layer, the depth to which the sediment is influenced by exchanging with the overlying water column. In addition, the thickness in the model formulation should characterize a reasonable time history or memory in the sediment layer. It is known that the temporal and spatial scales are much smaller in the sediment system than the water column. As such, the memory of the sediment layer is much longer than the water column. However, too thin a sediment layer will remember or be influenced by deposition of material that would have occurred only within the last year or two of the period being analyzed. If the layer used in the model is too thick, the model will average too long a history, not reflecting substantial reductions in sedimentary phosphorus resulting from reduced phosphorus discharges from wastewater treatment plants. Two recent studies[23,58] on the sediments in the Chesapeake Bay and Patuxent Estuary offered some important guidance on selecting the sediment depth(s) for the model. For example, the material accumulated in 5 cm of the sediment column in the Patuxent represents between 5 and 20 years of deposition, depending on the location in the bay. In this study, the sediment layer depths together with the assigned sedimentation velocities would provide for a ten year detention time or memory which is a reasonable approximation of the active sediment layer in the Patuxent.

A. MODELING APPROACH

The reactions which convert algal and refractory carbon to their end products are complex. The algal and refractory carbon are first converted to reactive intermediates and the process is similar to the refractory organic and algal nitrogen degradation. As such, the rates for carbon and nitrogen decomposition are assumed to be equal.[59] However, the reactive intermediates participate in further reactions: for example, volatile acids react to become methane, and the mechanisms that control these reactions are somewhat uncertain. In addition, few measurements of these intermediate species are available. Thus, a simplified yet realistic formulation of these reactions which was first developed by Di Toro and Connolly[60] and adopted by Thomann and Fitzpatrick[59] for the Potomac Estuary model, was employed for the Patuxent Estuary model.[57]

The approach was based on separating the initial reactions that convert sedimentary organic material into reactive intermediates, and the remaining redox reactions that occur. Then using a transformation variable and an orthogonality relationship, the mass balance equations, independent of the details of the redox equations, can be derived. As a result, the details of the redox equations can be avoided.[59] Instead, functions of the component concentrations are used to compute only the component concentrations, which can be treated in the same way as any other variable in the mass transport calculation. The convenient choice of components for the calculation are those that parallel the water quality variables in the water column: CBOD and DO. Restricting the calculation to these components, however, eliminates the possibility of explicitly including the effects of other reduced species, such as iron, manganese, and sulfide which play a role in overall redox reactions and may be involved in the generation of sediment oxygen demand. The decomposition reactions which drive the component mass balance equations are the anaerobic decomposition of the algal carbon, and the anaerobic breakdown of the sedimentary organic carbon. Both reactions are sinks of the oxygen and rapidly drive its concentration negative, indicating that the sediment is reduced rather than oxidized. The negative concentrations computed can be thought of as the oxygen equivalents of the reduced end products produced by the chains of redox reactions occurring in the sediment. Since the calculated concentration of oxygen is positive in the overlying water, it is assumed that the reduced carbon species (negative oxygen equivalents) that are transported across the sediment-water interface combine with the available oxygen and are oxidized to CO_2 and H_2O with a consequent reduction of oxygen in the overlying water column.

In the Patuxent sediment system, detrital algae and zooplankton are decomposed to produce ammonia and organic nitrogen. Particulate organic nitrogen is hydrolyzed to ammonia by bacterial action within the sediment. In addition to the ammonia produced by the hydrolysis of particulate organic nitrogen in the sediment, ammonia is generated by the anaerobic decomposition of algae. However, the end product is not exclusively ammonia, at least initially. Rather, a fraction of the algal nitrogen becomes particulate organic nitrogen, which must undergo hydrolysis before becoming ammonia. Ammonia produced by the hydrolysis of nonalgal organic nitrogen and the decomposition of detrital algal nitrogen may then be fluxed to the overlying water column via diffusive exchange. Organic nitrogen is hydrolyzed to ammonia in a temperature-dependent reaction.

Ammonia is lost from the sediment through diffusion into the overlying water column. No nitrification occurs in the sediment due to the anaerobic conditions present in the sediment. However, denitrification, the conversion of nitrate to nitrogen gas, may occur. Nitrate is present in the sediment due to diffusive exchange with the overlying water. The formulation of the sediment nitrogen concentrations and the resulting flux of ammonia are relatively straightforward because of the simplicity of the kinetics: hydrolysis and anaerobic algal decay produce a stable end product, ammonia, which does not undergo further reactions in the anaerobic sediment.

The anaerobic decomposition of algae and zooplankton in the sediments produces phosphorus in three forms: dissolved inorganic phosphorus (DIP), dissolved organic phosphorus (DOP), and particulate organic phosphorus (POP). For the phosphorus components, mineralization of organic phosphorus to inorganic forms is assumed to occur using the same rate expressions and rate constants as for organic nitrogen. A fraction of the end product, inorganic phosphorus, remains in the interstitial water, with the remainder being involved in the formation of precipitates or being sorbed onto the sediment solids. The fraction of DIP in the interstitial water varies spatially according to the sediment concentrations and the partitioning coefficient in the

TABLE 5
Sediment Layer $CBOD_5$ and DO Reactions

$CBOD_5$	$S_{CBOD} = a_{oc} k_{pzd} \theta_{pzd}^{T-20} P_c / BOD_{u5} - k_{ds} \theta_{ds}^{T-20} CBOD_u - 1.25 a_{on} k_{120} \theta^{T-20} NO_3$
DO:	$S_{DO} = -k_{ds} \theta_{ds}^{T-20} CBOD_5 BOD_{u5}$
Sediment Oxygen Demand (SOD):	$SOD = \frac{\xi}{h}([DO]_{wc} - [DO]_{sed})$

where

k_{pzd} = anaerobic algal decomposition rate (d^{-1})
θ_{pzd} = temperature coefficient
k_{ds} = organic carbon (as CBOD) decomposition rate (d^{-1})
θ_{ds} = temperature coefficient
ξ = molecular diffusion coefficient across the sediment-water interface (cm^2/sec)
a_{oc} = oxygen to carbon ratio (2.67)
a_{on} = oxygen to nitrogen ratio (2.29)
P_c = phytoplankton biomass (mg C/l)
BOD_{u5} = ultimate to 5-d BOD ratio
h = sediment layer depth
$[\,]_{sed}$ = sediment concentration
$[\,]_{wc}$ = water column concentration

TABLE 6
Sediment Layer Nitrogen Reactions

Total Organic Nitrogen (TON):	$S_{TON} = a_{nc} f_{on} \left(k_{pzd} \theta_{pzd}^{T-20}\right)[P_c] - k_{ond} \theta_{ond}^{T-20} [TON]$
Ammonia Nitrogen (NH_3):	$S_{NH_3} = a_{nc} f_{NH_3} \left(k_{pzd} \theta_{pzd}^{T-20}\right)[P_c] + k_{ond} \theta_{ond}^{T-20} [TON]$
Nitrate Nitrogen (NO_3):	$S_{NO_3} = -k_{12} \theta_{12}^{T-20}$
Sediment Ammonia Flux:	$NH_{3_{flux}} = \frac{\xi}{h}([NH_3]_{sed} - [NH_3]_{wc})$
Sediment Nitrate Flux:	$NO_{3_{flux}} = \frac{\xi}{h}([NO_3]_{sed} - [NO_3]_{wc})$

where

a_{nc} = nitrogen to carbon ration
f_{on} = fraction of dead phytoplankton recycled to the organic nitrogen pool
k_{ond} = organic nitrogen decomposition rate (d^{-1})
θ_{ond} = temperature coefficient

sediment. Exchange of the dissolved phosphorus with the overlying water is also similar to that of ammonia, nitrate, and dissolved oxygen. It should be stressed that the effects of hypoxia, as observed in the lower Patuxent Estuary, upon sediment phosphorus flux are incorporated into the modeling framework using the two-layer water column configuration. The sediment layer model formulations for sediment oxygen demand, nitrogen and phosphorus reactions are summarized in Tables 5 to 7.

TABLE 7
Sediment Layer Phosphorus Reactions

Dissolved Organic Phosphorus (DOP): $S_{DOP} = a_{pc} f_{dop} \left(k_{pzd} \theta_{pzd}^{T-20} \right) [P_c] - k_{opd} \theta_{opd}^{T-20} [DOP]$

Particulate Organic Phosphorus (POP): $S_{POP} = a_{pc} f_{pop} \left(k_{pzd} \theta_{pzd}^{T-20} \right) [P_c] - k_{opd} \theta_{opd}^{T-20} [POP]$

Dissolved Inorganic Phosphorus (DIP): $S_{DIP} = a_{pc} f_{dip} \left(k_{pzd} \theta_{pzd}^{T-20} \right) [P_c] + k_{opd} \theta_{opd}^{T-20} ([DOP] + [POP])$

Particulate Inorganic Phosphorus (PIP): $S_{PIP} = 0$

Repartitioning Phosphorus: Total Inorganic Phosphorus (TIP)

$$TIP = DIP + PIP$$

$$PIP = f_{p_{sed}} TIP$$

$$DIP = \left(1 - f_{p_{sed}} TIP \right)$$

Sediment Phosphorus Flux: $P_{flux} = \frac{\xi}{h} ([DIP]_{sed} - [DIP]_{wc})$

where

a_{pc} = phosphorus to carbon ration
f_{dop} = fraction of dead phytoplankton recycled to dissolved organic phosphorus pool
f_{pop} = fraction of dead phytoplankton recycled to particulate organic phosphorus pool
f_{dip} = fraction of dead phytoplankton recycled to inorganic phosphorus pool
k_{opd} = organic phosphorus decomposition rate (d^{-1})
θ_{opd} = temperature coefficient
f_{sed}= fraction particulate in the sediment layer
h = depth of the sediment layer

B. MODEL RESULTS

The sediment oxygen demand, ortho-phosphate and ammonia fluxes across the sediment-water interface calculated by the model are plotted for comparison with the data. Figures 8 to 10 show the these quantities at three locations: Station XDE5339 (Broomes Island), Station XDE2792 (St. Leonard Creek), and Buena Vista, respectively over a three-year period (1983–1985). No flux data are available at Broomes Island for comparison with the model results. However, the seasonal trends of SOD and ammonia fluxes at Buena Vista are somewhat reproduced by the model results. The comparison at St. Leonard Creek is not as good as that for Buena Vista. Phosphorus fluxes calculated by the model display sharp peaks during the summer months when the bottom water becomes hypoxic. However, the phosphorus flux peaks are short-lived and less significant in 1983 as compared with those in 1984 and 1985, resulting in calculated inorganic phosphorus concentrations in the water column lower than the data during 1983 (see Chapter 4). These peaks are difficult to verify as the field measurements lack sufficient temporal resolution.

Longitudinal profiles of sediment oxygen demand and nutrient fluxes are presented in Figure 11 for four seasons in 1985. Again, data is very limited for a good quantitative comparison with the model results. Nevertheless, the spatial trends along the estuary are reasonable although the ammonia fluxes calculated by the model are consistently higher than the measured fluxes.

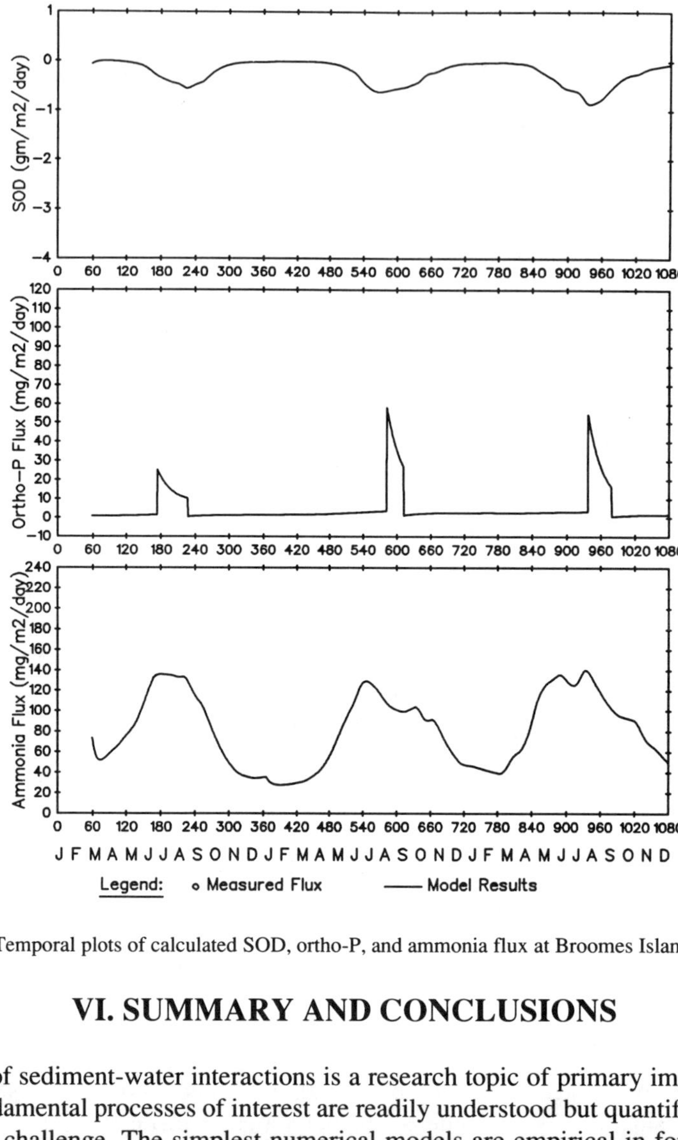

FIGURE 8. Temporal plots of calculated SOD, ortho-P, and ammonia flux at Broomes Island (1983–1985).

VI. SUMMARY AND CONCLUSIONS

Modeling of sediment-water interactions is a research topic of primary importance at this time. The fundamental processes of interest are readily understood but quantification of these processes is a challenge. The simplest numerical models are empirical in form. In terms of successfully reproducing observations, empirical models compare favorably with more complex mechanistic models but empirical models are not suited to address the effects of altering POM input to the sediments. A steady-state diagenetic model, based on a formulation employed in the study of marine sediment biogeochemistry, has been introduced. The model is valuable in illustrating the nonlinear dependence of SOD on organic input and the reciprocal relationship between SOD and COD export to the water column. The model illustrates the coupling of sediment nitrification and denitrification and offers insights into the importance of denitrification as a permanent nitrogen sink. Substantial nitrate release and enhanced anoxic phosphorus release are not reproduced, however, because they are transient rather than steady-state processes.

Estuaries are rarely in steady state. Water quality problems such as algal blooms and summer anoxia are among the processes of primary interest in the examination of eutrophication. Inevitably, time-variable sediment models must be applied to lend insight into the interactions of sediment-water fluxes and seasonal variations in estuarine characteristics. The Patuxent Estuary water quality model is a good example. In fact, long-term model simulations would provide key information for developing a sound water quality management strategy for the estuary.

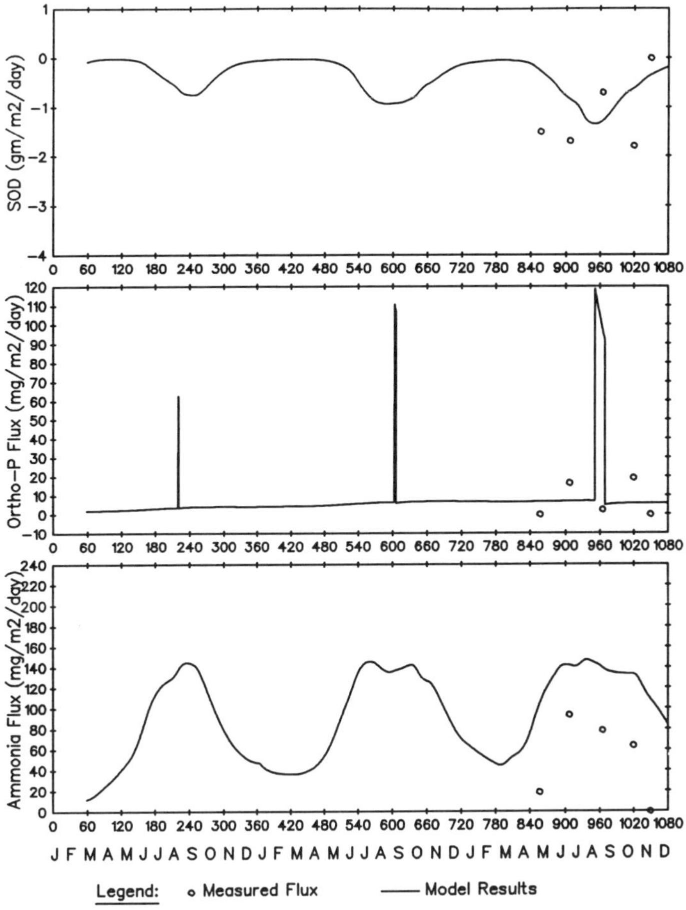

FIGURE 9. Temporal plots of calculated SOD, ortho-P, and ammonia flux at St. Leonard Creek (1983–1985).

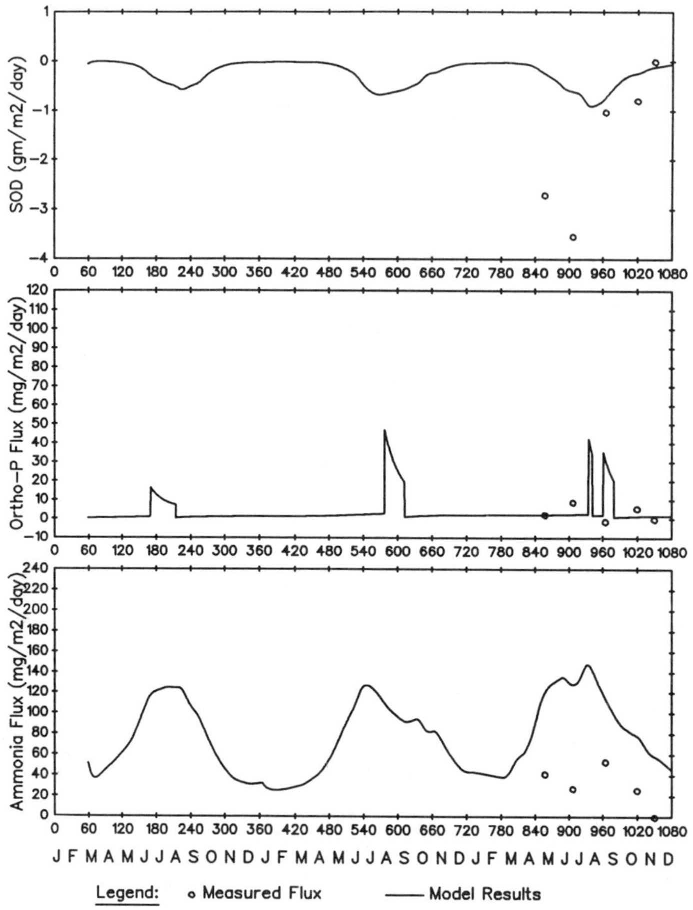

FIGURE 10. Temporal plots of calculated SOD, ortho-P, and ammonia flux at Buena Vista (1983–1985).

REFERENCES

1. **HydroQual, Inc.,** A Steady-State Coupled Hydrodynamic/Water Quality Model of the Eutrophication and Anoxia Process in Chesapeake Bay, report prepared for U.S. Environmental Protection Agency, Annapolis, MD, 1987.
2. **O'Connor, D., Gallagher, T., and Hallden, J.,** Water Quality Analysis of the Patuxent River, report prepared for U.S. Environmental Protection Agency and State of Maryland Department of Health and Mental Hygiene by HydroQual Inc., Mahwah, NJ, 1981.
3. **Thomann, R., Jaworski, N., Nixon, S., Paerl, H., and Taft, J.,** The 1983 Algal Bloom in the Potomac Estuary, Potomac Strategy State/EPA Management Committee, U.S. Environmental Protection Agency Region III, Philadelphia, 1985.
4. **HydroQual, Inc.,** Water Quality Analysis of the James and Appomattox Rivers, report prepared for Richmond Regional Planning District Commission, 1986.
5. **Berner, R.,** *Early Diagenesis*, Princeton University Press, Princeton, 1980, chaps. 1–3.
6. **Stumm, W. and Morgan, J. J.,** *Aquatic Chemistry*, 2nd ed., John Wiley & Sons, New York, 1981, 7.
7. **Di Toro, D. M.,** A diagenetic oxygen equivalents model of sediment oxygen demand, in *Sediment Oxygen Demand*, Hatcher, K., Ed., Institute of Natural Resources, University of Georgia, Athens, 1986, 171.
8. **Di Toro, D. M., Paquin, P., Subburamu, S., and Gruber, D.,** Sediment oxygen demand: methane and ammonia oxidation, *J. Environ. Eng.*, April 1988, submitted.
9. **Wang, W.,** Fractionation of sediment oxygen demand, *Water Res.*, 14, 603, 1980.
10. **Pamatmat, M.,** Oxygen consumption by the seabed. IV. Shipboard and laboratory experiments, *Limnol. Oceanogr.*, 16, 536, 1971.

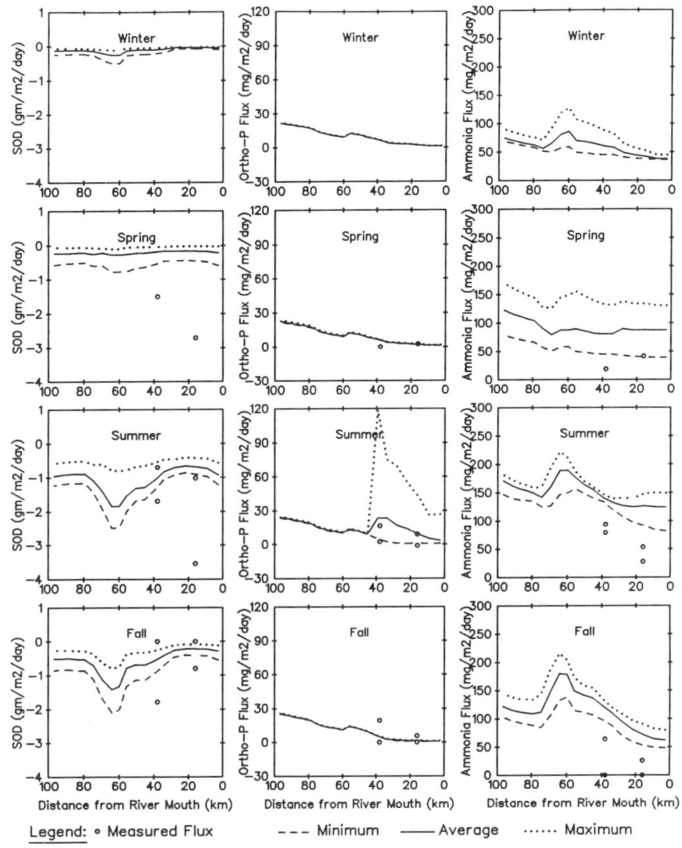

FIGURE 11. Spatial plots of calculated SOD, ortho-P, and ammonia flux in 1985.

11. **Brewer, W., Abernathy, A., and Paynter, M.,** Oxygen consumption by freshwater sediments, *Water Res.*, 11, 471, 1977.
12. **Smith, K., Burns, K., and Teal, J.,** In situ respiration of benthic communities of Castle Harbor, Bermuda, *Mar. Biol.*, 12, 196, 1972.
13. **Hale, S.,** The role of benthic communities in the nitrogen and phosphorus cycles of an estuary, in *Mineral Cycling in Southeastern Ecosystems*, Howell, F., Ed., ERDA Symposium Series, University of Rhode Island Sea Grant Advisory Service, Narragansett, 1975, 291.
14. **Boynton, W., Kemp, W. M., and Osborne, C.,** Nutrient fluxes across the sediment-water interface in the turbid zone of a coastal plain estuary, in *Estuarine Perspectives*, Kennedy, V., Ed., Academic Press, New York, 1980, 93.
15. **Simon, N. and Kennedy, M.,** The distribution of nitrogen species and adsorption of ammonium in sediments from the tidal Potomac River and estuary, *Estuar. Coast. Shelf Sci.*, 25, 11, 1987.
16. **Nixon, S.,** Remineralization and nutrient cycling in coastal marine ecosystems, in *Estuaries and Nutrients*, Neilson, B. and Cronin, L., Eds., Humana Press, Clifton NJ, 1981, 111.
17. **Krom, M. and Berner, R.,** Adsorption of phosphate in anoxic marine sediments, *Limnol. Oceanogr.*, 25, 797, 1980.
18. **Cerco, C.,** Sediment-Water Column Exchanges of Nutrients and Dissolved Oxygen in the Tidal James and Appomattox Rivers, report PB85-242915 XAB, National Technical Information Service, Springfield, VA, 1985.
19. **Fillos, J. and Molof, A.,** Effect of benthol deposits on oxygen and nutrient economy of flowing waters, *Journal Water Pollut. Cont. Fed.*, 44, 644, 1972.
20. **Sorensen, J.,** Reduction of ferric iron in anaerobic marine sediment and interaction with reduction of nitrate and sulfate, *Appl. Environ. Microbiol.*, 43, 319, 1982.
21. **Lijklema, L.,** The role of iron in the exchange of phosphate between water and sediments, in *Interactions Between Sediments and Freshwater*, Golterman, H., Ed., Dr. W. Junk B.V., The Hague, 1977, 313.
22. **Cerco, C.,** Sediment nutrient fluxes in a tidal freshwater embayment, *Water Res. Bull.*, 24, 255, 1988.

23. **Boynton, W. and Kemp, W. M.,** Nutrient regeneration and oxygen consumption by sediments along an estuarine salinity gradient, *Mar. Ecol. Prog. Ser.*, 23, 45, 1985.
24. **Cerco, C.,** Measured and modelled effects of temperature, dissolved oxygen, and nutrient concentration on sediment-water nutrient exchange, *Hydrobiologia*, in press.
25. **Fisher, T., Carlson, P., and Barber, R.,** Sediment nutrient regeneration in three North Carolina estuaries, *Estuar. Coast. Shelf Sci.*, 14, 101, 1982.
26. **Callender, E. and Hammond, D.,** Nutrient exchange across the sediment-water interface in the Potomac River estuary, *Estuar. Coast. Shelf Sci.*, 15, 395, 1982.
27. **Bowden, W.,** A nitrogen-15 isotope dilution study of ammonium production and consumption in a marsh sediment, *Limnol. Oceanogr.*, 29, 1004, 1984.
28. **Van Kessel, J.,** Factors affecting the denitrification rate in two water-sediment systems, *Water Res.*, 11, 259, 1977.
29. **Messer, J. and Brezonik, P.,** Laboratory evaluation of kinetic parameters for lake sediment denitrification models, *Ecol. Model.*, 21, 277, 1984.
30. **Nedwell, D.,** Exchange of nitrate, and the products of bacterial decomposition between seawater and sediment from a U.K. saltmarsh, *Estuar. Coast. Shelf. Sci.*, 14, 557, 1982.
31. **Jenkins, M. and Kemp, W. M.,** The coupling of nitrification and denitrification in two estuarine sediments, *Limnol. Oceanogr.*, 29, 609, 1984.
32. **Seitzinger, S., Nixon, S., and Pilson, M.,** Denitrification and nitrous oxide production in a coastal marine ecosystem, *Limnol. Oceanogr.*, 29, 73, 1984.
33. **Klapwijk, A. and Snodgrass, W.,** Experimental measurement of sediment nitrification and denitrification in Hamilton Harbour, Canada, *Hydrobiologia*, 91, 207, 1982.
34. **Pomeroy, L., Smith, E., and Grant, C.,** The exchange of phosphate between estuarine water and sediments, *Limnol. Oceanogr.*, 10, 167, 1965.
35. **Fillos, J. and Swanson, W.,** The release rate of nutrients from river and lake sediments, *J. Water Pollut. Cont. Fed.*, 44, 1032, 1975.
36. **Holdren, G. and Armstrong, D.,** Factors affecting phosphorus release from intact lake sediment cores, *Environ. Sci. Technol.*, 14, 79, 1980.
37. **Nixon, S., Kelly, J., Furnas, B., Oviatt, C., and Hale, S.,** Phosphorus regeneration and the metabolism of coastal marine bottom communities, in *Marine Benthic Dynamics*, Tenare, K. and Coull, B., Eds., University of South Carolina Press, Columbia, 1980, 219.
38. **Edberg, N. and Hofsten, B.,** Oxygen uptake of bottom sediments studied in situ and in the laboratory, *Water Res.*, 7, 1285, 1973.
39. **Edwards, R. and Rolley, H.,** Oxygen consumption of river muds, *J. Ecol.*, 53, 1, 1965.
40. **Cerco, C.,** Effect of Temperature and Dissolved Oxygen on Sediment-Water Nutrient Flux, report PB86-151982 lA06, National Technical Information Service, Springfield, VA, 1985.
41. **Vanderborght, J. and Billen, G.,** Vertical distribution of nitrate concentration in interstitial water of marine sediments with nitrification and denitrification, *Limnol. Oceanogr.*, 20, 953, 1975.
42. **Vanderborght, J., Wollast, R., and Billen, G.,** Kinetic models of diagenesis in disturbed sediments. I. Mass transfer properties and silica diagenesis, *Limnol. Oceanogr.*, 22, 787, 1977.
43. **Galloway, F. and Bender, M.,** Diagenetic models of interstitial nitrate profiles in deep sea suboxic sediments, *Limnol. Oceanogr.*, 27, 624, 1982.
44. **Jahnke, R., Emerson, S., and Murray, J.,** A model of oxygen reduction, denitrification, and organic matter mineralization in sediments, *Limnol. Oceanogr.*, 27, 610, 1982.
45. **Berner, R.,** Stoichiometric models for nutrient regeneration in anoxic sediments, *Limnol. Oceanogr.*, 22, 781, 1977.
46. **HydroQual, Inc.,** Evaluation of Sediment Oxygen Demand in the Upper Potomac Estuary, report prepared for Metropolitan Washington Council of Governments, Washington, D.C., 1987.
47. **Malone, T., Kemp, W. M., Ducklow, H., Boynton, W., Tuttle, J., and Jonas, R.,** Lateral variation in the production and fate of phytoplankton in a partially stratified estuary, *Mar. Ecol. Prog. Ser.*, 32, 149, 1986.
48. **Barcelona, M.,** Sediment oxygen demand fractionation kinetics, and reduced chemical substances, *Water Res.*, 17, 1081, 1983.
49. **Almgren, T. and Hagstrom, I.,** The oxidation rate of sulfide in seawater, *Water Res.*, 8, 395, 1973.
50. **Painter, H.,** A review on literature on inorganic nitrogen metabolism in microorganisms, *Water Res.*, 4, 393, 1970.
51. **McCutcheon, S. C.,** Laboratory and instream nitrification rates for selected streams, *J. Environ. Eng.*, 113, 628, 1987.
52. **Kristensen, E. and Blackburn, T.,** The fate of organic carbon and nitrogen in experimental marine sediment systems: influence of bioturbation and anoxia, *J. Mar. Res.*, 45, 231, 1987.
53. **Schubel, J. R. and Pritchard, D.,** Responses of upper Chesapeake Bay to variations in discharge of the Susquehanna River, *Estuaries*, 9, 236, 1986.

54. **Lauria, J. and Goodman, A.,** Measurement of sediment interstitial water COD gradient for estimating the sediment oxygen demand, in *Sediment Oxygen Demand*, Hatcher, K., Ed., Institute of Natural Resources, University of Georgia, Athens, 1986, 367.
55. **Campbell, P. and Rigler, F.,** Effect of ambient oxygen concentration on measurements of sediment oxygen consumption, *Can. J. Fish. Aquat.*
56. **Seitzinger, S.,** Nitrogen biogeochemistry in an unpolluted estuary: the importance of benthic denitrification, *Mar. Ecol. Prog. Ser.*, 41, 177, 1987.
57. **Lung, W. S.,** A Water Quality Model for the Patuxent Estuary, final report submitted to Maryland Department of the Environment, University of Virginia, Department of Civil Engineering, Environmental Engineering Research Report No. 8, 1992.
58. **Boynton, W. R., Kemp, W. M., Garber, J. H., Barnes J. M., Cowan, J. L. W., Stammerjohn, S. E., Matteson, L. L., Rphland, F. M., and Marvin, M.,** Long-Term Characteristics and Trends of Benthic Oxygen and Nutrient Fluxes in the Maryland Portion of Chesapeake Bay, paper presented at the 1990 Chesapeake Bay Conference, Baltimore, MD.
59. **Thomann, R. V. and Fitzpatrick, J. J.,** Calibration and Verification of the Potomac Estuary Model, HydroQual, Inc., final report prepared for the Washington, D.C. Department of Environmental Services, 1982.
60. **Di Toro, D. M. and Connolly, J. P.,** Mathematical Models of Water Quality in Large Lakes, Part 2: Lake Erie, EPA-600/3-80-065, 1980.

Chapter 7

INTEGRATING HYDRODYNAMIC AND WATER QUALITY MODELS

I. INTRODUCTION

Over the past two decades, estuarine modeling has advanced along a dual course (i.e., hydrodynamic modeling and water quality modeling) with little coupling of water quantity and water quality calculations. Previously, its was considered sufficient to assume steady flow, complete mixing, and simple dilution in estuarine water quality modeling. Attention was focused on the kinetics of reaeration, biodegradation, and the like, as in the classic case of oxygen sag; hydrodynamics per se was not the concern of the analysts. The observation by Orlob,[1] over 20 years ago provides us with a perspective on this topic:

> Success in verifying water quality models has generally been less than satisfactory, largely because of the paucity of good data, but also because of the greater dependence on empiricism in structuring the models. As noted previously, the relative dependence on the advection and diffusion terms in the mass transport equation is a function of the scale of the model. As the scale increases, i.e., segments become larger or phenomena are averaged over longer time steps, the dependence on the empirical effective diffusion coefficient increases. Verification of coarse scale mathematical models is usually accomplished by subjective adjustment of the coefficient until the model prediction agrees with prototype performance. Because the coefficients are derived from historic experience with the prototype, they are usually not considered reliable for prediction of prototype performance under conditions that differ markedly from the historic.

Little advancement to such a modeling approach has been made in the 15 years following Orlob's observation—at least until a few years ago.

Estuarine flow and physical transport processes are characterized by a wide range of temporal and spatial scales, as well as significant nonlinear interactions associated with both tidal and buoyancy dynamics. Temporal scales may range from the daily and monthly scales associated with tidal phenomena, to the seasonal and yearly scales associated with river discharge or buoyancy variability. Biogeochemical transformation processes in estuarine systems are also characterized by a variety of temporal and spatial scales. The coupled nutrient and biological cycles exhibit daily, seasonal, and yearly time scales, while nutrient loading and local concentration and biological species predominance covers spatial scales ranging over typical vertical to horizontal physical dimensions. Toxic contaminants, particularly those interacting with sediments, may have residence times of decades.

The hierarchy of spatially reduced models, including one-dimensional longitudinal, two-dimensional vertical and horizontal and quasi-three-dimensional (i.e., two-dimensional horizontal coupled with one-dimensional vertical) is well established. Such transport models are formally based upon advection-dispersion equations with transport processes associated with the unresolved spatial dimension(s) parameterized by longitudinal dispersion coefficients or dispersion coefficient tensors.

Given the spatial and temporal variability of estuarine hydrodynamic, transport and biogeochemical processes, three-dimensional time-variable models are apparently the most desirable as management tools. However, technical and institutional constraints, as well as model adequacy, pose serious obstacles to their use, particularly if long-term simulations are required.

Thus, the common approach used to quantify mass transport for water quality modeling is primarily an empirical one. In many cases, the dispersion coefficient is determined by comparing a solution of the mass transfer equation with the measured concentration distribution of some substance in an estuary. The same dispersion coefficient is then used to predict the concentration distribution of some other substance. In practice, the empirical determination of dispersion coefficients is limited by the requirement that all source and sink (decay) terms for the substance must be known with reasonable accuracy. Therefore, this method is restricted to conservative substances, such as salinity, or to artificially introduced tracers for which decay and adsorption rates can be independently determined. In general, observed data from either the prototype or a hydraulic model is used to calibrate the dispersion coefficient. This methodology is relatively straightforward and widely used. However, back-calculating mass transport could limit the model's credibility since a large degree of freedom can be used to adjust transport coefficients to fit the salinity data. Such exercises often result in mass transport models which are overly empirical having dispersion and turbulent diffusion parameterizations with little or no physical basis as observed by Orlob.[1] It is difficult to obtain a unique transport pattern, associated with a given salinity distribution, in a one-dimensional modeling framework, let alone two- or three-dimensional calculations. This drawback is particularly obvious when velocity data is lacking to independently check the advective transport. Furthermore, the derived transport pattern(s) would not be valid for a different hydrologic and hydrodynamic condition. As a result, predictive capability is limited as the methodology relies on given distribution of salinity or conservative substances.

The above discussion demonstrates why many estuarine water quality models, with state-of-the-art kinetics to describe biological processes, lack the hydrodynamic component and as a result, lack full predictive capability. Increasing interest in fate and transport modeling of pollutants, including toxic substances, has provided the necessary stimulus to improve predictive capability through the coupling of hydrodynamic solutions and water quality calculations.

One of the most recent case studies is the water quality modeling of the Chesapeake Bay for eutrophication control. The modeling strategy formulated called for the phased development of a three-dimensional, time-variable water quality model. Initial efforts were focused on a steady-state model that would assess the general response of the bay to nutrient loadings and would identify the important processes to be included in the time-variable effort. An important finding from the steady-state model was that sediment fluxes play a key role in the water quality of the bay. Due to the relatively slow reaction rates of diagenetic processes in the sediment, which recycle inorganic nutrients to the water column, it appears as if the modeling framework must be able to perform long-term simulations in order to assess water column and sediment response to nutrient control strategies. Long term water quality simulations require the use of hydrodynamic models to quantify the current patterns and distributions of diffusion/dispersion throughout the bay. Given the rather intensive computational requirements needed to solve multidimensional hydrodynamic and water quality models, it is a significant effort to interface hydrodynamic and water quality models so that long term simulations are practical. The next question is how should a hydrodynamic model be linked with a water quality model for water quality problem(s) in a given estuary?

The purpose of this chapter is to report some of the recent development in estuarine mass transport models which are based on the hydrodynamic model results. This mass transport model would serve as the linkage between the hydrodynamic model and the water quality model.

II. DIFFERENT TEMPORAL MODES IN MASS TRANSPORT REPRESENTATION

Estuarine mass transport may be represented in a number of temporal modes:

- Intratidal real-time mass transport
- Tidally averaged mass transport under time-variable conditions
- Slack water approximation under time-variable conditions

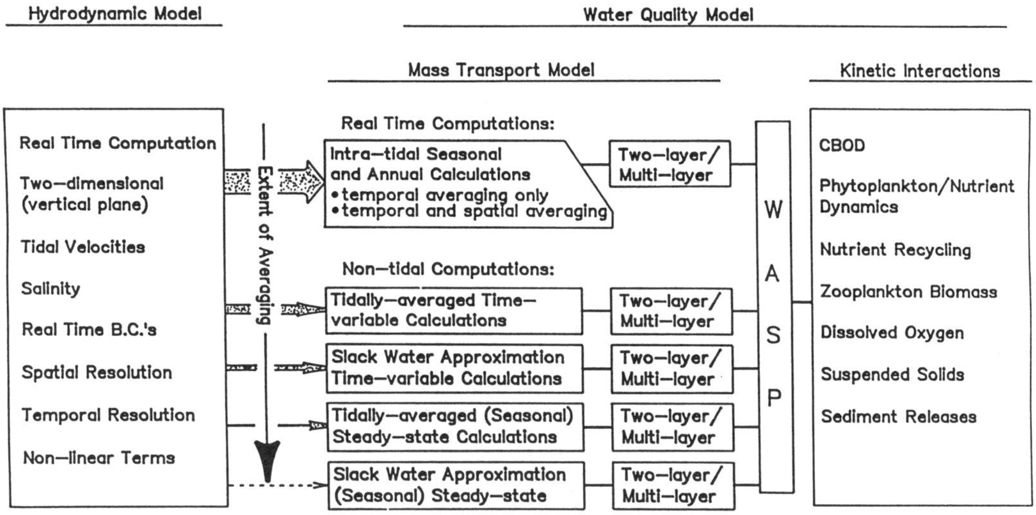

FIGURE 1. Temporal averaging schemes to integrate two-dimensional hydrodynamic and water quality models.

- Tidally averaged mass transport under steady-state conditions
- Slack water approximation under steady-state conditions

The first category is real-time, intratidal computations while the last four categories are nontidal computations. Additional insight into the relationship between hydrodynamic solutions, mass transport calculations, and water quality models, is displayed in Figure 1. A two-dimensional laterally averaged hydrodynamic model is used for this illustration. As indicated, the linkage between the hydrodynamic and mass transport models are strong in real-time computations. The linkage is progressively decreasing in nontidal computations as the extent of temporal averaging increases. In fact, the slack water approximation for seasonal steady-state calculations usually does not require hydrodynamic solutions as the empirical approach of matching the salinity distribution is employed.

The effort required for real-time or time-variable mass transport analysis is resource demanding and computational intensive, often resulting in complex models. In addition, data requirements to calibrate and verify the model are formidable. The data on the hour-to-hour changes in freshwater inflow, point source discharges, and nonpoint source input must be available for model calibration in addition to the tidal stage and velocity data on an hourly basis. Such a database is rarely available in many applications. Yet, time-variable modeling results are sometimes time-averaged in wasteload allocations

In eutrophication modeling, tidally averaged mass transport may be considered, recognizing that minute-to-minute or hour-to-hour variation is not appropriate for algal growth dynamics. It is argued that all of the inputs that are relevant to the growth dynamics cannot be specified on so fine a time scale, and the phytoplankton population does not significantly vary from hour to hour. Thus, tidally averaged mass transport is appropriate when seasonal water quality changes (such as dissolved oxygen) associated with eutrophication are more important than diurnal fluctuations. In this case, water quality surveys must be conducted throughout a season or an entire year to reflect the variabilities in tides, freshwater flow rates, and water quality conditions.

Further, most estuaries can be sampled at slack water just at the turn or the end of the flood/ebb tide. The next water sample can be taken at each successive upstream station at slack water and a synoptic picture is obtained of the variation of the water quality at a fixed stage of the tide.[2] Recent modeling studies using slack water quality survey data include the Patuxent Estuary[3] and the Potomac Estuary.[4]

Again, the drawback of these temporal reduced mass transport models is they lack the predictive capability. A compromised approach is to link the hydrodynamic and water quality models using a mass transport modeling scheme. While this mass transport scheme will allow temporal averaging, it should also retain as much hydrodynamic results as possible.

III. TECHNICAL CONSIDERATIONS

It is necessary to review a few technical aspects of the coupling and their difficulties in estuarine modeling. The review provides physical insight into the problems, and offers a perspective to coupling estuarine hydrodynamic and water quality models.

A. REAL-TIME VS. NONTIDAL COMPUTATIONS

This issue has received much attention from many researchers in estuarine water quality computations and from those using estuarine models. In fact, continuing debates between these two approaches have existed for many years. One may explore the conditions under which it is necessary and appropriate to utilize a hydrodynamic model to calculate, for example, the hour-to-hour tidal currents and stage. There are two principal reasons for calculating water motions via a hydrodynamic model[5]

- There is uncertainty about the hydrodynamic circulation and motion, as might be the case in a complicated interconnected coastal bay and harbor system or in an estuary under a transient input of freshwater flow.
- Some structural change is anticipated (e.g., dike, dredging, diversion) that requires a prediction of a new hydrodynamic regime or circulation prior to a determination of the impact of the proposed change on water quality.

In the case of the Potomac Estuary, and in light of the eutrophication issues, a hydrodynamic calculation appears appropriate only at the short-term time scale and local to intermediate spatial scales, i.e., for problems associated with combined sewer overflow and transient nonpoint source impacts. At longer time and large space scales appropriate to seasonal dissolved oxygen questions and the eutrophication/nutrient issue, there is no substantive uncertainty about the magnitude or direction of the water flow. As a result, those water quality issues for the Potomac Estuary involving seasonal and longer term time scales, and regional to comprehensive spatial scales do not require a prior hydrodynamic intratidal calculation.[5]

On the other hand, Harleman[3] suggested that the time scale associated with the water quality model is not necessarily determined by the time scale of interest in a water quality problem. The small time scale of the real-time models should be considered as a necessary solution technique which avoids the ambiguous coefficients associated with the nontidal advective models. The choice is again one of a trade-off between the time and expense devoted to analysis as opposed to the time and expense of the additional field work, such as special dye dispersion tests, required by the large time scale, nontidal water quality models.[3]

It is clear that there will continue to be a place for the simpler nontidal mass transport models in estuarine water quality simulations. For example, both the real-time and nontidal models could be formulated initially. Solutions for both models could be obtained for a single effluent source undergoing a simple first-order decay. The real-time model could then be used to verify the hydrodynamic aspects of the nontidal model through the choice of nontidal dispersion coefficients to a desired degree of agreement with the real-time solution. Thus, the formulation and solution of the real-time problem is an "analytical dye dispersion test" which replaces the prototype or hydraulic model "dye dispersion test", which replaces the prototype, or hydraulic "dye dispersion test" previously used to verify the nontidal model.

B. TEMPORAL AVERAGING AND SPATIAL REDUCTION OF HYDRODYNAMIC RESULTS

During the model development phase for the Patuxent Estuary,[7] a coarse grid box mass transport model, based on a methodology by Lung and O'Connor,[8] was used to calibrate the water column kinetics in the estuary. The results from a 38-segment model which determines the two-layer mass transport by matching the calculated salinity distributions with observed data is shown in Figure 11 in Chapter 3. Such an ad hoc determination of the mass transport is not suitable for long-term simulations when salinity distributions are not available *a priori*. Thus, a hydrodynamic model, which provides the mass transport information for the water quality model, is needed. More importantly, the hydrodynamic model must be linked with the water quality model in a proper and efficient manner. In general, the time scales associated with the water quality processes are much greater than those for tidal currents and stages. Further, the spatial scales for the water quality model can also be considerably larger than those of the hydrodynamic model. Thus, it may not be prudent to link the two models using the same temporal and spatial resolutions. Temporal and/or spatial averaging of the the hydrodynamic model results are therefore desirable. On the other hand, excessive averaging may lose much information from the hydrodynamic model, resulting in unrealistic mass transport coefficients.

C. DISPERSION

Temporal reduction and spatial averaging contribute to apparent dispersion. For example, the major component of the one-dimensional longitudinal dispersion may be due to the vertical variation of the horizontal velocity. To date, dispersion represents the largest difficulty in water quality modeling, particularly using the box model. Shanahan and Harleman[9] analyzed the difficulties in the appropriate use of the dispersion term in the box model formulation and raised objections to the use of the box model in many applications. Two major problems arise, in regard to dispersion, in the application of the box model. First, it is difficult to quantify an appropriate dispersion coefficient for use in the box model. Second, the box model may be numerically overdispersive unless relatively small segments are used. In laterally averaged two-dimensional box models for estuaries, increase in spatial resolution in the vertical direction would reduce longitudinal dispersion because characterization of mixing is improved.

D. NUMERICAL DISPERSION

This issue has been addressed briefly in Chapter 3. In general, appropriate numerical schemes are used in the hydrodynamic model to reduce or eliminate numerical dispersion. When applying box models to water quality simulations, numerical dispersion is inherent. For example, using the program WASP always has certain amounts of numerical dispersion.[10] When calculating vertical concentration gradients (e.g., salinity) in partially mixed estuaries, excessive numerical dispersion in the water quality model could result in overly flat concentration gradients even though the associated hydrodynamic calculations match the vertical gradient nicely. In time-variable calculations, excessive numerical dispersion in box models may be partially controlled by reducing the box size or by selecting time steps which are sufficiently large to approach the Courant limit.[11]

E. RATE CONSTANTS FOR BIOCHEMICAL TRANSFORMATION

Estuarine water quality models usually contain a number of state variables and rate constants. Since a water quality model coupled with a hydrodynamic model is usually calibrated (and sometimes validated) against a set of data, the question often raised is to what extent are the rate constants attributed to biochemical processes influenced by the segmentation and mass transport of the model through coupling. Our ability to represent complex aquatic biochemical transformation by means of a mathematical model is less well developed than the state of knowledge in the geophysical area. Thus, it is essential that the possibility of spurious influences resulting from improper coupling of hydrodynamic and biochemical submodels be recognized, and that the effect of this interrelationship be quantified.

IV. TEMPORAL AVERAGING OF HYDRODYNAMIC MODEL RESULTS

Lung and Hwang[12] developed a mass transport model that averages the hydrodynamic model results over a period significantly longer than the integration step of the hydrodynamic model for the Patuxent Estuary. The Patuxent Estuary hydrodynamic model is a two-dimensional, laterally averaged model that is based on the Wang-Kravitz[13] modeling framework. The mass transport model is a box model using WASP (see Chapter 4) which is the basis of the water quality modeling computations. The key test in such a model interfacing is comparing the real-time salinity results from the hydrodynamic model with the salinity results from the mass transport (box) model on a time-variable basis. This comparison is straightforward using the dense to dense grid linking approach. That is, this is simply a time-filtering exercise without any reduction of spatial resolution.

A. THE PATUXENT ESTUARY HYDRODYNAMIC MODEL

Olson and Kincaid[14] developed the Patuxent Estuary hydrodynamic model, which is based on the laterally-averaged shallow water equations of motion. In their model, the momentum balance in the vertical direction is hydrostatic. The governing equations are conservation of mass, horizontal momentum and salinity, plus a linear equation of state.

$$\frac{\partial(uB)}{\partial x} + \frac{\partial(wB)}{\partial z} = 0 \tag{7.1}$$

$$\frac{\partial(uB)}{\partial t} + \frac{\partial(uuB)}{\partial x} + \frac{\partial(uwB)}{\partial z} + gB\frac{\partial\eta}{\partial x} = -\frac{gB}{\rho}\int_0^z \frac{\partial p}{\partial x} + \frac{\partial}{\partial x}\left[\frac{BN_x \partial u}{\partial x}\right] + \frac{\partial}{\partial z}\left[\frac{BN_z \partial u}{\partial z}\right] - C_D u|u|\frac{\partial B}{\partial z} \tag{7.2}$$

$$\frac{1}{\rho}\frac{\partial p}{\partial z} = g \tag{7.3}$$

$$\frac{\partial SB}{\partial t} + \frac{\partial(SuB)}{\partial x} + \frac{\partial(SwB)}{\partial z} = \frac{\partial}{\partial x}\left[\frac{BD_x \partial S}{\partial x}\right] + \frac{\partial}{\partial z}\left[\frac{BD_z \partial S}{\partial z}\right] \tag{7.4}$$

$$\rho = \rho_o(1 + \beta S) \tag{7.5}$$

where

t = time
p = pressure
g = gravitational acceleration
N_x and N_z = eddy diffusivity for momentum
u and w = mean velocity components
x and z = rectangular coordinates

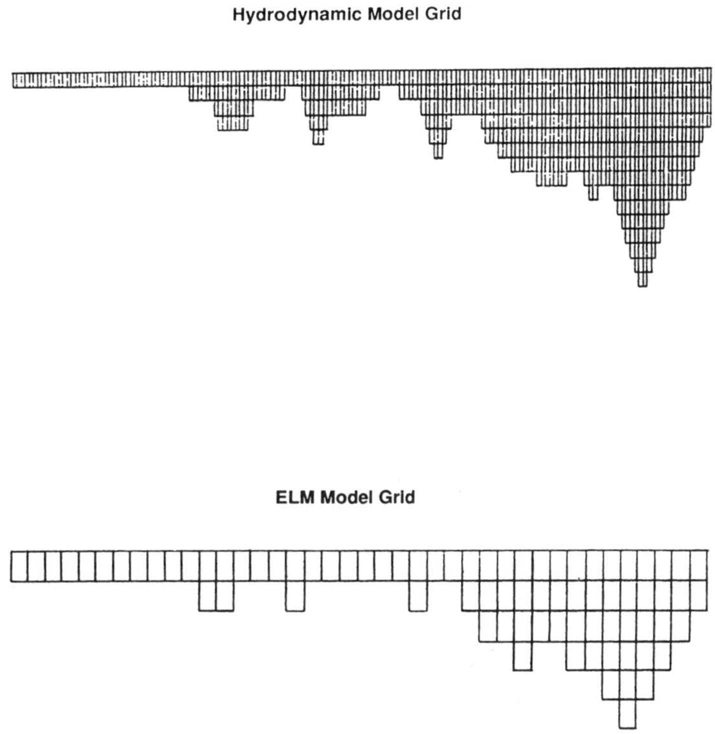

FIGURE 2. Two-dimensional hydrodynamic model grid for the Patuxent Estuary.

C_D = bottom drag coefficient = 1.6×10^{-3} (dimensionless)
D_x and D_z = turbulent diffusion coefficients for mass
S = salinity (ppt)
ρ = density
ρ_o = density of freshwater
β = 7.29×10^{-4} (ppt^{-1})

The boundary conditions used by Olson and Kincaid[14] are at the channel bottom; the normal velocity and salt flux both vanish. In addition, the bottom stress is characterized by the bottom velocity. At the water surface, the vertical velocity and salinity gradient both vanish. The upstream boundary condition is specified by the freshwater inflow. The salt concentration in the freshwater inflow can be ignored. At the downstream boundary where the Patuxent joins the main stem of the Chesapeake Bay, the longitudinal gradient of the advective transport is zero. The tidal stage is specified with the measured data.

Olson and Kincaid[14] used spatially and temporally constant eddy diffusivity and turbulent diffusion coefficients as follows:

D_x = 1.25×10^5 cm^2/sec
D_z = 0.20 cm^2/sec
N_x = 2.00×10^7 cm^2/sec
N_z = 5.00 cm^2/sec

The governing equations are solved on a two-dimensional finite difference grid system with spatial intervals of 2m in the vertical direction and 638 m in the longitudinal direction (Figure 2). A semi-implicit method by Wang and Kravitz[13] was used as the time-stepping procedure with a time step of one minute for the Patuxent Estuary.

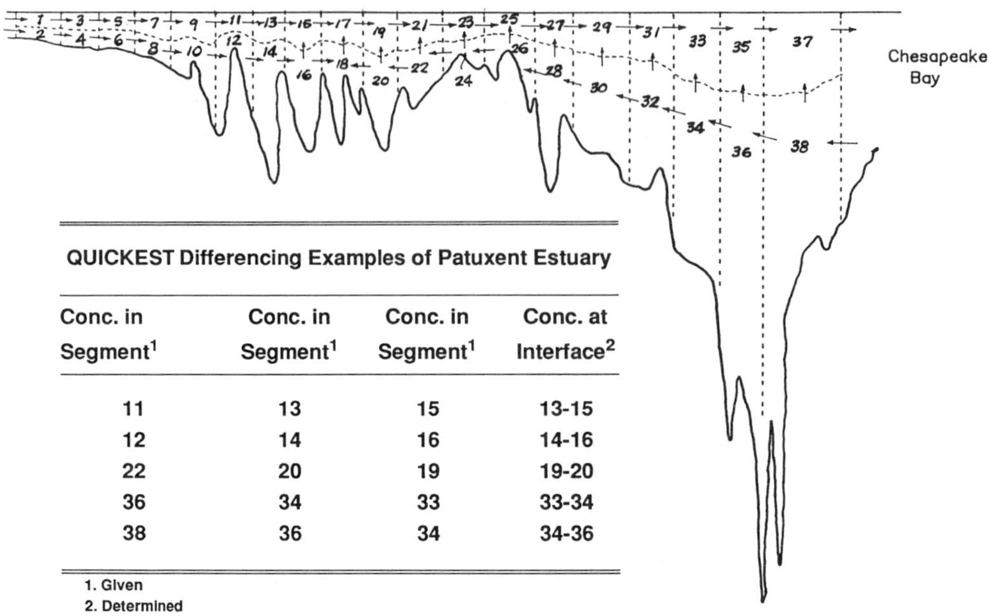

QUICKEST Differencing Examples of Patuxent Estuary

Conc. in Segment[1]	Conc. in Segment[1]	Conc. in Segment[1]	Conc. at Interface[2]
11	13	15	13-15
12	14	16	14-16
22	20	19	19-20
36	34	33	33-34
38	36	34	34-36

1. Given
2. Determined

FIGURE 3. QUICKEST scheme.

The hydrodynamic model has been calibrated and verified with field data collected during two intensive surveys: spring and fall 1986. The model results reproduced the observed pattern of salinity and current velocity during those periods to within an acceptable error range.[14]

B. MASS TRANSPORT CALCULATIONS — SALINITY

In this calculation, the same spatial configuration as the hydrodynamic model is used for mass transport calculations. Based on the same grid in the hydrodynamic model, the mass transport model consists of a total 873 segments with 165 segments in the longitudinal direction and 2 to 16 segments in the vertical direction, depending on the depth of the channel. Temporal averaging of velocities in the hydrodynamic model is made for use in the water quality model.

The numerical computations face two constraints: time step and numerical dispersion, which are closely related to each other. The upwind differencing scheme used in the box model to guarantee positive solutions yields numerical dispersion. The magnitude of numerical dispersion is a function of the Courant number. Given the same spatial and temporal resolution as in the hydrodynamic model, the mass transport computations in the water quality model cannot be carried out without generating excessive numerical dispersion. While longer time steps will increase the Courant number and thereby reduce numerical dispersion to some extent, there is an upper bound for the time step to maintain numerical stability.

To reduce numerical dispersion, the QUICKEST scheme developed by Leonard[15] was adopted for the two-dimensional box model in this study. Additional discussion of the numerical scheme is also found in Abbott and Basco.[16] In QUICKEST, the concentration at a segment interface is approximated by a quadratic interpolation using the concentrations in the two adjacent segments and in the next upstream segment (Figure 3). This approximation yields a third-order accuracy in space. Numerical boundaries, suggested by McBride,[17] were used at the physical boundaries to avoid artificially generating sources or sinks of mass. That is, the wall concentration and concentration gradient of the boundary cell were interpolated or extrapolated using the concentrations at the last two interior segments. The quadratic interpolation function is the same as that used for interior segments. It should be pointed out that the QUICKEST scheme presented by Leonard[15] and by Abbott and Basco[16] is for constant grid size in finite

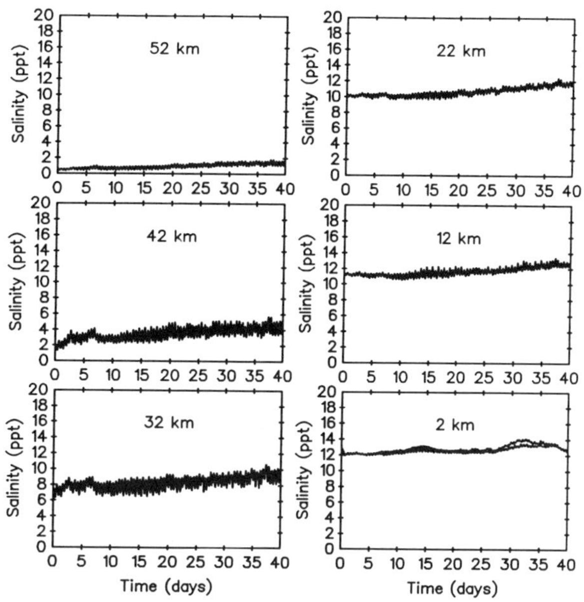

FIGURE 4. Comparing salinity results between hydrodynamic and temporally averaged mass transport models for the Patuxent Estuary, spring 1986.

difference applications. In the Patuxent Estuary study,[12] the segments are in different sizes in the salinity mass transport model; yet, the finite difference scheme is used in the numerical solution of the mass balance equation for salinity.

C. MASS TRANSPORT MODEL RESULTS

A crucial test of the mass transport computations using the hydrodynamic model results is to compare the calculated salinity with that generated by the hydrodynamic model. Such a test was conducted for two different periods in 1986: spring and fall intensive surveys by Olson and Kincaid.[13]

With the box model configuration the initial mass transport computations used the hydrodynamic time step. That is, a direct link-up of the hydrodynamic model and the water quality model was performed. Then, the averaging period was progressively increased to 15 and 30 hydrodynamic time steps. A comparison between the 1- and 30-min time step calculations had revealed insignificant differences in salinity distributions. This outcome implies that the temporal averaging of velocity over 30 min does not cause significant change in dispersion coefficient. It should be pointed out that it is simpler to use the same spatial grid for the hydrodynamic and the water quality box model. Using the same spatial grid avoids spatial averaging problems when attempting to deal with temporal averaging.

Leonard[15] indicated that under the QUICKEST scheme, the Courant number could be as high as 2.0 without causing numerical instability. With a time step of 30 min in this calculation, the Courant numbers are generally in the range of 0.5 to 1.5. Further increase of the time step is not possible because of this constraint. Nevertheless, the QUICKEST scheme not only reduces numerical dispersion, it also doubled the box model time step from 15 to 30 min.

Figure 4 presents the salinity results from the hydrodynamic (1-min time step) and the mass transport/water quality (30-min time step) models at the 4-m depth throughout the estuary for the period from May 7 to June 15, 1986. The water quality model mimics the hydrodynamic model closely. The relatively significant tidal fluctuations of salinity at segments 60 and 80 are due to strong salinity variations during tidal excursions in midestuary where the longitudinal salinity gradient reaches maximum.

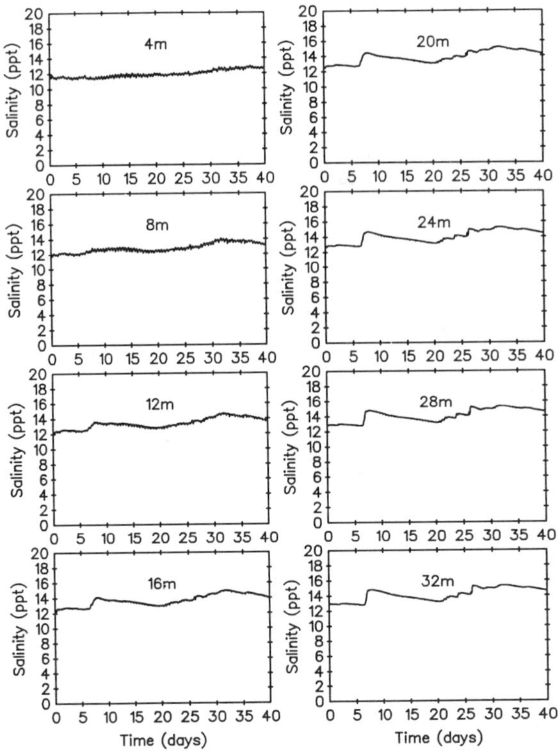

FIGURE 5. Comparing salinity results at various depths in the lower estuary between hydrodynamic and mass transport models.

Salinity results at a 4-m interval in the deepest section of the estuary for the same period are summarized in Figure 5. Again, the two salinity calculations match each other very well throughout the water column. At about 160 h, a salt front enters the estuary from the Chesapeake Bay, resulting in accumulation of salt in the deep pockets of the Patuxent Estuary.

Different transport schemes were used for transport calculations in these two models, resulting in different dispersion coefficients. Thus, the dispersion coefficients obtained through calibration in the QUICKEST scheme were several times of those used in the upwind scheme of the hydrodynamic model, indicating reduced numerical dispersion using QUICKEST.

It is seen that a simple averaging procedure was used in this analysis to process the hydrodynamic model results for water quality model calculations of the Patuxent Estuary. Technical difficulties such as time step size and numerical dispersion were encountered in the box model configuration. The QUICKEST scheme was used to reduce numerical dispersion and to increase the time step. The maximum time step used for the mass transport calculations in the water quality model is 30 min, which is much longer than the hydrodynamic calculation time step (1 min). The 30-min time step is limited by the fine spatial resolution of the grid and further increases in the step size would make the computation unstable.

V. SPATIAL AVERAGING OF HYDRODYNAMIC MODEL RESULTS

It is quite obvious that in order to capitalize on the ability of the hydrodynamic model to calculate mass transport (advection and mixing), the averaging period for interfacing must be considerably less than the tidal period (30 min was the maximum as shown previously).

Numerical dispersion and integration time steps are two factors against each other in this type of Eulerian transport model. It appears that there is an interesting trade-off between the spatial resolution and the computational time step for the water quality model. In other words, there may be an advantage in using the maximum amount of information about the mass transport from the hydrodynamic linkage.

A. EULERIAN-LAGRANGIAN APPROACH

The Eulerian-Lagrangian transport models offer a good alternative to the box type, Eulerian transport models. One of the important possibilities provided by this alternate approach is that considerably longer time steps (even beyond a tidal period) may be used in the water quality model.

While the Eulerian approach treats the concentrations as a function of time and location in fixed coordinates, the Lagrangian approach follows the movement of each water parcel, and treats the concentrations as a function of time and their previous locations. To take advantage of the Lagrangian treatment of advective transport while avoiding its difficulties in coordinates deformation, the Eulerian-Lagrangian Method (ELM) was introduced by Leith.[18] The ELM traces the movement of water parcels on a fixed Eulerian computational framework. The trajectories of water parcels and the constituent concentrations are calculated for each grid point at each computational time step. Since the last location of a water parcel at the end of each time step is generally not a grid point, interpolation is required to approximate the concentration.

The averaging of flow velocity and constituent concentration over space in a model coupling can be avoided by using an Eulerian-Lagrangian approach. The Eulerian-Lagrangian Method traces the movement of water parcels on a fixed Eulerian computational framework. The trajectories of water parcels and the constituent concentrations are calculated for each grid point at each computational time step. Since the final position of a water parcel at the end of each simulation time step is generally not a nodal point, interpolation is required to approximate the concentration for this point using the known nodal values. Using this method, transport computations can be carried out at desired locations without spatially averaging flow velocity. Thus, the difficulties accrue from spatial average of flow velocity can be avoided.

A two-dimensional Lagrangian-Eulerian transport model can be derived from a two-dimensional transport equation. Equation 7.6 represents a two-dimensional advection-diffusion transport,

$$\frac{\partial \bar{c}}{\partial t} = \bar{u}\frac{\partial \bar{c}}{\partial x} + \bar{w}\frac{\partial \bar{c}}{\partial z} + D_x\frac{\partial^2 \bar{c}}{\partial x^2} + D_z\frac{\partial^2 \bar{c}}{\partial z^2} \tag{7.6}$$

Considering advective transport only, the equation can be represented using a total derivative D/Dt in a trajectory defined by $\bar{u} = dx/dt$ and $\bar{w} = dz/dt$,

$$\frac{D\bar{c}}{Dt} = 0 \tag{7.7}$$

Thus, the concentration at a point Q at time $(n + 1)\mathrm{D}t$ can be considered to be equal to the concentration at time $n\mathrm{D}t$ at a point P, which, under the specified velocity field, reaches point Q at time $(n + 1)\mathrm{D}t$, as shown in Figure 6. Hence, the calculation of advection is replaced by the task of first integrating the velocities through $\mathrm{D}t$ to find the previous position of point P, and then estimating the concentration at this point, c_p^n, from the known concentrations at surrounding nodal points, $c_{i,j}^n$. This scheme is computationally stable. Its accuracy relies on the method used for concentration interpolation. Interpolation using a normalized second-order Lagrangian polynomial has been proven to be accurate and free of numerical dispersion by Cheng et al.[19] In this method, the concentration at a point P is calculated from its nine surrounding nodal points as shown in Figure 7, using

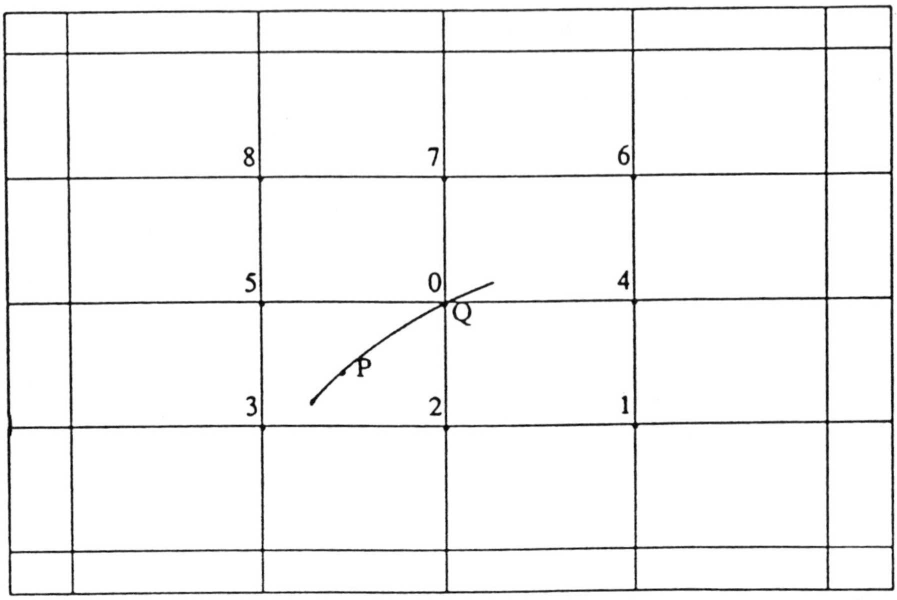

FIGURE 6. LaGrangian method scheme.

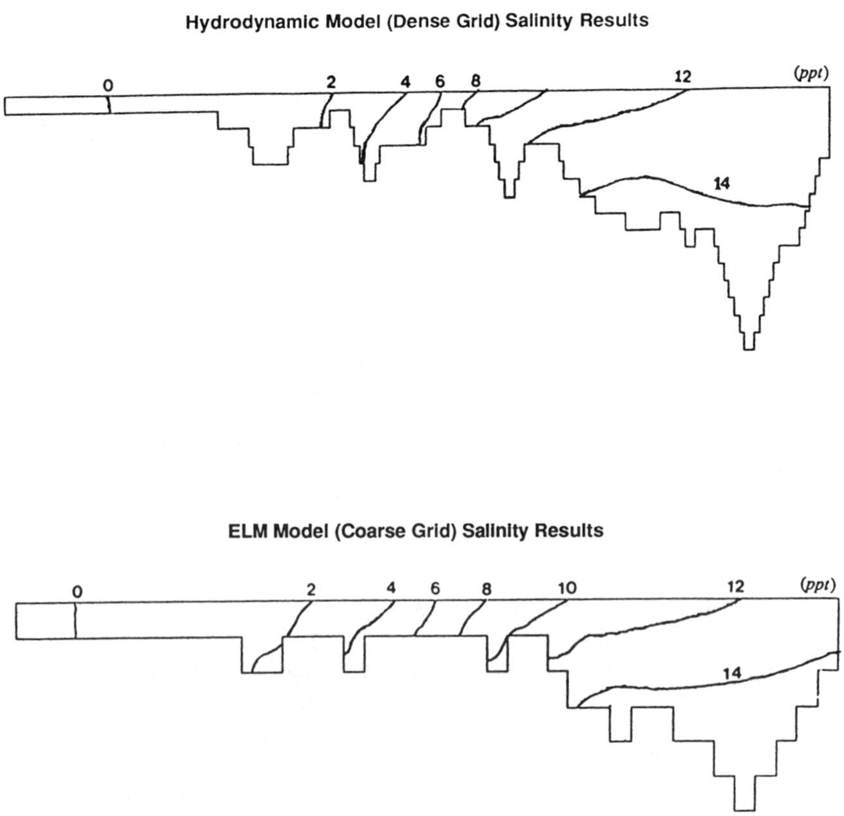

FIGURE 7. Comparing two-dimensional salinity distributions between coarse and dense grid mass transport model results using ELM, at the end of 40-d simulation.

$$c_P^n = \sum_{k=0}^{8} N_k(\zeta,\eta)c_k^n \tag{7.8}$$

where z and h are the normalized position function in local coordinates. Their values are calculated from

$$\zeta = -u_{i,j}^{n+1}\frac{\Delta t}{\Delta x} \tag{7.9}$$

and

$$\eta = -w_{i,j}^{n+1}\frac{\Delta t}{\Delta z} \tag{7.10}$$

The interpolation functions N_k(z, h)are similar to the shape functions used in a finite element scheme. Their values are calculated from

$$\begin{aligned}
N_0 &= (1-\zeta^2)(1-\eta^2)\\
N_1 &= -\frac{1}{4}\zeta(1+\zeta)\eta(1-\eta)\\
N_2 &= -\frac{1}{2}(1-\zeta^2)\eta(1-\eta)\\
N_3 &= \frac{1}{4}\zeta(1-\zeta)\eta(1-\eta)\\
N_4 &= \frac{1}{2}\zeta(1+\zeta)(1-\eta^2)\\
N_5 &= -\frac{1}{2}\zeta(1-\zeta)(1-\eta^2)\\
N_6 &= \frac{1}{4}\zeta(1+\zeta)\eta(1+\eta)\\
N_7 &= \frac{1}{2}(1-\zeta^2)\eta(1+\eta)\\
N_8 &= -\frac{1}{4}\zeta(1-\zeta)\eta(1+\eta)
\end{aligned} \tag{7.11}$$

Test results by Hwang and Lung[20] showed that this scheme is both accurate and stable. In that test, a concentration step function is subject to a tidal fluctuation. Using a time step of 15 min and simulating over a period of 10 d, the computation is stable. The Lagrangian feature of this method facilitates the flexibility of choosing locations where concentrations are to be calculated. Therefore, a coarse computational grid can be established by using a subset of the original dense grid. Spatial averaging of flow velocity, as in a pure Eulerian approach, is not necessary. Thus, the difficulty of redefining mixing coefficients is avoided. This feature makes the ELM ideal for coupling a dense grid hydrodynamic model with a coarse grid water quality transport model. The streak-line of a water parcel under a coarse grid system will still be the same as it is under a dense grid system. However, the model results using coarse segmentation will not be identical to those using dense segmentation because, under coarse segmentation, the interpolation is done using concentrations at coarse grid points instead of those on the dense grid points. Again, how close the coarse grid model results are to the dense grid model results depends largely on how well the interpolation is done. The second-order Lagrangian polynomials just shown can adequately describe the concentration profiles in most natural water bodies.

B. DIFFUSIVE TRANSPORT SIMULATION

The diffusive part of a laterally-averaged two-dimensional transport equation is

$$B\frac{\partial \bar{c}}{\partial t} = \frac{\partial}{\partial x}\left(BD_x\frac{\partial \bar{c}}{\partial x}\right) + \frac{\partial}{\partial z}\left(BD_z\frac{\partial \bar{c}}{\partial z}\right) \tag{7.12}$$

The mixing coefficients D_x and D_z include the mechanisms of molecular diffusion, turbulent diffusion and dispersion. They are assumed to be constants in space and time. Quite a few methods are available for the numerical solution of this equation.[21] To be compatible with the ELM advective transport simulation, it is desirable that the numerical scheme for diffusive transport also be relatively stable and computationally simple. Under such considerations, a DuFort-Frankel method is considered to be ideal for dispersive transport simulation.[22] The difference form of Equation 7.12, using this scheme is

$$\begin{aligned} B_{i,j}\frac{c_{i,j}^{n+1} - c_{i,j}^{n-1}}{2\Delta t} &= D_x\left(\frac{B_{i+1/2,j}\left(c_{i+1,j}^{n} - c_{i,j}^{n-1}\right) - B_{i-1/2,j}\left(c_{i,j}^{n+1} - c_{i-1,j}^{n}\right)}{\Delta x^2}\right) \\ &+ D_z\left(\frac{B_{i,j+1/2}\left(c_{i,j+1}^{n} - c_{i,j}^{n-1}\right) - B_{i,j-1/2}\left(c_{i,j}^{n+1} - c_{i,j-1}^{n}\right)}{\Delta z^2}\right) \end{aligned} \tag{7.13}$$

This scheme is unconditionally stable. It requires more computational effort than other frequently used schemes because concentrations at three time levels are needed. Otherwise, the computation is relatively simple because iterations are not necessary. The scheme has a first-order accuracy. Although more complicated schemes, such as the Crank-Nicolson method and other implicit schemes, yield more accurate results, application of these schemes is much more costly in terms of computer time.[23] In an advective system, the accuracy of a DuFort-Frankel method will be adequate for diffusive transport simulations.[20]

C. APPLICATION TO THE PATUXENT ESTUARY

1. Spatial Reduction Using ELM

The ELM method was applied to the Patuxent Estuary to develop a mass transport model with a reduced spatial resolution from the hydrodynamic model grid. The ELM method was first applied to the hydrodynamic calculation using the original dense grid (873 elements) and then applied to a coarse grid (126 elements) under the spring 1986 condition. The two-dimensional salinity distributions calculated from both models are compared in Figure 7. The comparison indicates small differences in salinity distributions between these two model results. Figure 8 shows another comparison of salinity concentrations at 4-m depth for five locations along the estuary for a 40-d simulation from May 5th to June 15th, 1986. In addition, salinity results at three different depths of the location about 5 km above the river mouth are also displayed for comparison in Figure 8. The result indicates that spatial resolution reduction can be achieved using a Lagrangian approach without losing model predictive capability. Combining spatial and temporal reductions, the coupled model requires computational efforts only approximately 1/720 that of a direct link of the models (i.e., no reduction of spatial resolution). Use of the coarse computational grid also reduces computer memory requirements to one sixth that of the original dense grid model.

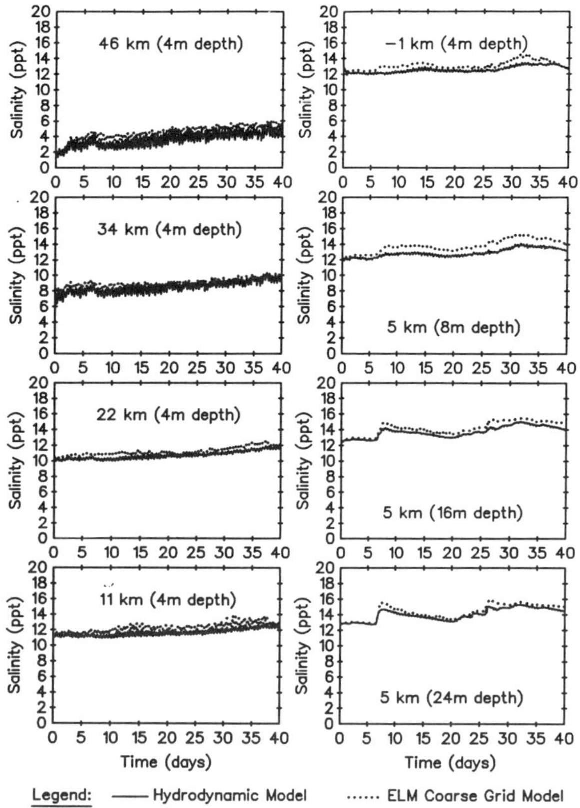

FIGURE 8. Comparing temporal salinity results between coarse and dense grid models using ELM, along the estuary at 4-m depth.

2. Model Simulations for Fall 1986 Condition

A separate set of data was used to verify the coupling schemes. The event of fall 1986 (from September 2nd to September 30th) represents a lower river flow, higher salinity concentration scenario. The magnitudes of salt diffusivity and momentum diffusivity used in the hydrodynamic simulation are identical to those for the spring simulation. The mixing coefficients for transport model computations were also identical to those used for the spring simulation. Results of these simulations are shown in Figure 9, showing the coarse grid model results matching the dense grid hydrodynamic model results very well.

3. Tidally Averaged Simulations

The tidally averaged simulation was based on the Lagrangian residual current as described previously. Results of this simulation are compared with the hydrodynamic model results. As shown in Figure 10, the tidal fluctuations of concentration disappear in an intertidal simulation. Nevertheless, the tidally averaged model successfully simulated the long-term trend of salinity distribution. A time lag in salinity variations is found in the tidally averaged model results. The time lag is a result of the long simulation time step used in the intertidal simulations. In a mathematical model, the constituent concentration is updated at discrete computational time steps. In reality, both advective and diffusive transport are continuous processes. Therefore, the sensitivity of the model to the concentration variations will be dependent on the computational

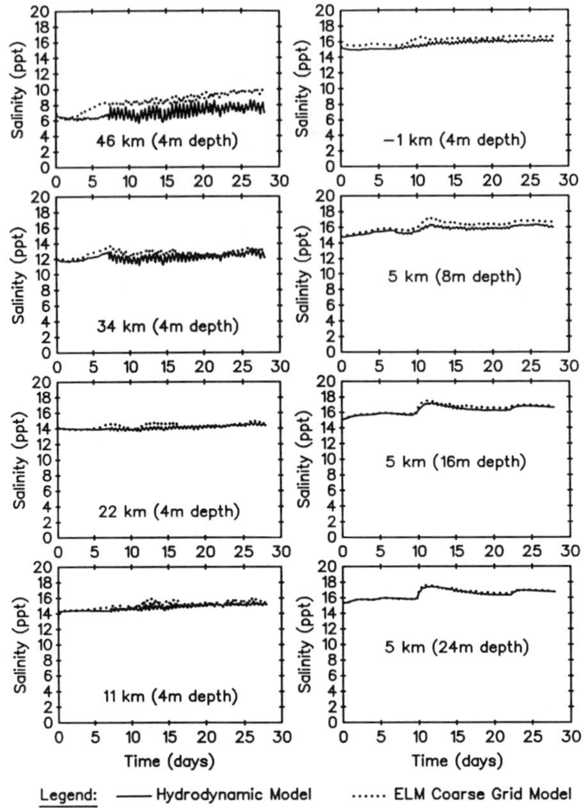

FIGURE 9. Salinity results from the coarse and dense grid models using ELM for fall 1986 condition.

time step used. The results of this simulation indicates that the tidally averaged model not only eliminates the tidal fluctuations of constituent concentration, it also has a damping effect on the concentration fluctuations at frequencies lower than the tidal cycle. Using coarse computational grid and a time step of 12.5 h, the model run requires only 1/4500 the time steps of the hydrodynamic model.

4. Discussions

The results of this study indicate that the computational efficiency and model predictive capability do not necessarily compromise one another. It is possible to develop a coupling scheme that is both efficient and predictive. It also confirms the advantage of a Lagrangian treatment of advective transport. A Lagrangian approach is superior to an Eulerian approach in that (1) a Lagrangian simulation of advective transport is computationally stable, and (2) since the nonphysical representation of advective transport is not necessary, diffusive transport can be accurately treated in a Lagrangian model.

Temporal integration is straightforward and more efficient than spatial integration. In the Patuxent Estuary, as well as other similar estuaries, the hydrodynamic information can be averaged over a period of approximately two hours without causing much loss of transport information. Therefore, temporal integration should be the initial consideration in a model coupling.

Results from the Patuxent Estuary study also indicated that the Lagrangian residual current provides adequate flow information for tidally averaged simulations. Using Lagrangian residual currents, tidally-averaged simulations can be carried out without redefining the diffusive coefficients.

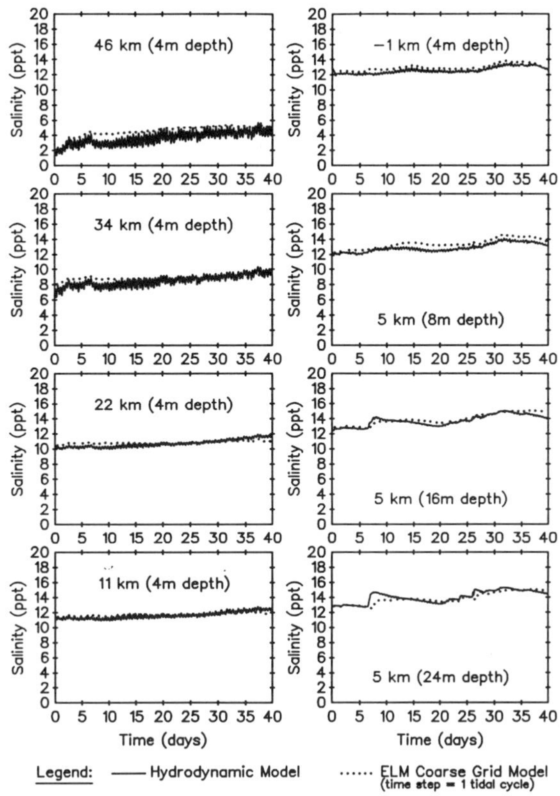

FIGURE 10. Salinity results from the hydrodynamic model and tidally-averaged coarse grid model.

VI. SUMMARY AND CONCLUSIONS

For long-term transport and transformation modeling, consistent with multiyear planning horizons, time-averaged or temporally filtered model equations are often used. Although temporal filtering is desirable from an economics of computation point of view, nonrigorous filtering has resulted in transport models that are overly empirical, having dispersion and turbulent diffusion parameterizations with little or no physical basis. Research has shown that the two-dimensional, in-the-vertical-plane, advection-dispersion equation may be rigorously filtered to remove tidal time scale fluctuations.

In the case study of the Patuxent Estuary, temporal averaging of the hydrodynamic results has been achieved for the mass transport model with integration steps up to 30 time steps of the hydrodynamic model. Successful spatial reduction scheme using the Eulerian-Lagrangian method has shown that integration steps of one tidal cycle is achievable for the Patuxent Estuary. It is envisioned that such an effort could result in water quality model time steps that are considerably longer than a tidal period. It appears that there is a delicate balance between the spatial resolution and the computational time step for the mass transport model used in water quality simulations. The overriding factor is, of course, to retain the most information from the hydrodynamic model.

In the last few years, other studies of integrating estuarine hydrodynamic and water quality models have been reported. For example, Dortch et al.[24] successfully linked a three-dimensional hydrodynamic model and a water quality model for the Chesapeake Bay. In their study, first-order Lagrangian residual currents are used to transport the water quality constituents. The effort was limited to temporal averaging of the hydrodynamic model results. No spatial reduction was attempted.

REFERENCES

1. **Orlob, G. T.,** Mathematical modeling of estuarial systems, in *Modeling of Water Resources Systems*, Biswas, A.K., Ed., Harvest House, 1972, 87.
2. **Thomann, R. V. and Mueller, J. A.,** *Principles of Surface Water Quality Modeling and Control*, Harper & Row, New York, 1987.
3. **Harleman, D. R. F.,** One-dimensional models, in *Estuarine Modeling: An Assessment*, TRACOR, Inc., prepared for the Office of Water Quality, U.S. Environmental Protection Agency, 1971, 3.
4. **Lung, W. S.,** *Developing a Water Quality Model for the Patuxent Estuary*, University of Virginia, Department of Civil Engineering, Environmental Engineering Technical Report No. 8, 1992.
5. **Thomann, R. V. and Fitzpatrick, J. J.,** Calibration and Verification of a Model of the Potomac Estuary, HydroQual, Inc., final report to Washington, D.C. Dept. of Environmental Services, 1982.
6. **Thomann, R. V.,** *Overview of Potomac Estuary Modeling, Tasks I and II — Dissolved Oxygen and Eutrophication*, HydroQual, Inc., 1980.
7. **Lung, W. S.,** Development of a water quality model for the Patuxent Estuary, in *Estuarine Water Quality Management*, Michaelis, W., Ed., Springer-Verlag, Berlin, Germany, 1990, 40.
8. **Lung, W. S. and O'Connor, D. J.,** Two-dimensional mass transport in estuaries, *J. Hydraulic Eng., ASCE*, 110, 1340, 1984.
9. **Shanahan, P. and Harleman, D. R. F.,** Transport in lake water quality modeling, *J. Environ. Eng.*, 110, 42, 1984.
10. **Di Toro, D. M., Fitzpatrick, J. J., and Thomann, R. V.,** Water Quality Analysis Simulation Program (WASP) and Model Verification Program (MVP) Documentation, U.S. Environmental Protection Agency, EPA-600/3-81-044, 1983.
11. **Bird, S., Hall, R., and Dortch, M.,** Coupling of hydrodynamic and water quality models, 1986.
12. **Lung, W. S. and Hwang, C. C.,** Integrating hydrodynamic and water quality models for Patuxent Estuary, in *Estuarine and Coastal Modeling*, ASCE, New York, 1989, 420.
13. **Wang, D. P. and Kravitz, D. W.,** A semi-implicit two-dimensional model of estuarine circulation, *J. Phys. Ocean.*, 10, 441, 1980.
14. **Olson, P. and Kincaid, C.,** Numerical model of Patuxent River Estuary hydrodynamics, The Johns Hopkins University, Geophysical Fluid Dynamics Laboratory technical report 89-1, 1991.
15. **Leonard, B. P.,** A stable and accurate convective modeling procedure based on quadratic upstream interpolation, *Comp. Meth. Appl. Mech. Eng.*, 19, 59, 1979.
16. **Abbott, M. B. and Basco, D. R.,** *Computational Fluid Dynamics An Introduction for Engineers*, Longman Scientific & Technical, New York, 1989, 5.
17. **McBride, G. B.,** QUICKEST algorithms for estuarine solute transport model, Hamilton Science Centre internal report no. IR/87/5, Ministry of Works and Development, Hamilton, New Zealand, 1987.
18. **Leith, C. E.,** Numerical simulation of the earth's atmosphere, *Meth. Comput. Phys.*, 4, 1, 1965.
19. **Cheng, R. T., Casulli, V., and Milford, S. N.,** Eulerian-Lagrangian solution of the convection-dispersion equation in natural coordinates, *Water Resour. Res.*, 20, 944, 1984.
20. **Hwang, C. C. and Lung, W. S.,** Coupling Estuarine Hydrodynamic and Water Quality Models, University of Virginia, Department of Civil Engineering, Environmental Engineering research report no. 3, 1990.
21. **Roache, P. J.,** *Computational Fluid Dynamics*, Hermosa Publishers, Albuquerque, NM, 1976.
22. **Richtmyer, R. D.,** *Difference Methods for Initial-Value Problems*, Interscience Publishers, Inc., New York, 1957.
23. **Delves, L. M.,** Techniques for large-scale problems, in *The Numerical Solution of Non-linear Problems*, Baker, C.T.H. and Phillips, C., Eds., Clarendon Press, Oxford, England, 1981, 300.
24. **Dortch, M. S., Chapman, R. S., Hamrick, J. M., and Gerald, T. K.,** Interfacing three-dimensional hydrodynamic and water quality models of Chesapeake Bay, in *Estuarine and Coastal Modeling*, Spaulding, M.L., Ed., ASCE, New York, 1990, 182.

Chapter 8

ESTUARINE MIXING ZONE ANALYSIS

I. INTRODUCTION

A. MIXING ZONE

In the water quality analysis presented so far, the wastewater effluent is considered to mix completely with the ambient water immediately at the outfall. In reality, a mixing zone is created in the vicinity of the outfall. The mixing zone is a region where complete mixing has not been achieved. When considering conventional pollutants, such as BOD and nutrients in estuaries, the spatial and temporal scales associated with this zone are insignificant compared with the characteristic spatial and temporal scales of the water quality problems associated with BOD and nutrients (see Figure 1, Chapter 2). As such, incomplete mixing is not an issue. However, when considering toxic contaminants, the high concentrations due to incomplete mixing could cause adverse impact on the aquatic biota even though the completely mixed concentrations meet the water quality standards in the receiving water.

With this understanding and when the water quality standards are adopted by the states, the standards are enforceable concentrations that apply to the indicated chemical species. Once a water quality standard is in place, it becomes necessary to establish concentrations (or effluent limitations) in the discharge stream that will prevent violations of the water quality standard. On the other hand, it is not always necessary to meet all water quality criteria within the discharge pipe to protect the integrity of the waterbody as a whole. Sometimes it is appropriate to allow for ambient concentrations above the criteria in small areas near outfalls. These areas are called mixing zones. Since these areas of impact, if disproportionately large, could potentially adversely impact the productivity of the water body, and have unanticipated ecological consequences, they should be carefully evaluated and appropriately limited in size.[1]

When the wastewater is discharged into the receiving water, its transport may be divided into two stages with distinctive mixing characteristics. Mixing and dilution in the first stage are determined by the initial momentum of the discharge. This initial contact with the receiving water is where the concentration of the effluent will be its greatest in the water column. The design of the discharge outfall should provide ample momentum to dilute the concentrations in the immediate contact area as quickly as possible. The second stage of mixing covers a more extensive area in which the effect of initial momentum is diminished and the waste is mixed primarily by ambient turbulence.

B. TOXICITY CRITERIA

To insure mixing zones do not impair the integrity of the estuary, it should be determined that the mixing zone will not cause lethality to passing organisms and, considering likely pathways of exposure, that there are no significant human health risks. For application of aquatic life criteria, there are two types of mixing zones. In the zone immediately surrounding the outfall, neither the acute nor the chronic criterion is met. The acute criterion is met at the edge of this zone. In the next mixing zone, the acute, but not the chronic, criterion is met.[2] The chronic criterion is met at the edge of the second mixing zone.

For toxic discharges, U.S. EPA[2] recommends careful evaluation of mixing to prevent zones of chronic toxicity that extend for excessive distances because of poor mixing. U.S. EPA maintains two water quality criteria for the allowable magnitude of toxic substances: a *criterion maximum concentration* (CMC) to protect against acute or lethal effects; and a *criterion continuous concentration* (CCC) to protect against chronic effects. Based on the discussions, it is understood that CMC should be met at the edge of the zone of initial dilution and CCC should be met at the edge of the overall mixing zone.

In rivers or tidal rivers that have a persistent throughflow in the downstream direction and do not exhibit significant natural density stratification, hydrologically based flows 1Q10 and 7Q10 for the CMC and CCC, respectively, have been used traditionally and now are recommended in steady-state mixing zone modeling analysis.[2]

II. MODELING ESTUARINE MIXING ZONES

A. DISCHARGE-INDUCED MIXING

Studies on a simple jet resulting from a discharge from a round nozzle into a body of water have been extensive. As a result, there is a good understanding of how they behave.[2] In practice, however, the discharge-induced mixing is quite complicated when buoyancy is involved. This type of mixing is probably more important in lakes and reservoirs and slow moving rivers since ambient mixing in those waterbodies is minimal.

In shallow environments, it is important to determine whether near-field instabilities occur. These instabilities, associated with surface and bottom interaction and localized recirculation cells extending over the entire water depth, can cause buildup of effluent concentrations by obstructing the effluent jet flow.[2] There are no simple means to estimate dilution in these cases. In the absence of near-field instabilities, horizontal or nearly horizontal discharges will create a clearly defined jet in the water column that will initially occupy only a small fraction of the available water depth.

For simple jets, the Gaussian distribution of tracer concentration across the jet is used to define the jet behavior:[3]

$$C = C_m \exp[-k(x/z)^2]$$

in which the subscript m refers to the value of C on the jet axis and where z is the distance along the jet axis and x is the transverse (or radial) distance from the jet axis.

A minimum estimate of the initial dilution available in the vicinity of a discharge can be made using the following equation derived from information in Holley and Jirka:[4]

$$S = 0.3\frac{x}{d}$$

in which S is flux-averaged dilution, x is distance from outlet, and d is diameter of outlet. The equation provides a minimum estimate of mixing because it is based on the assumptions that outlet velocity is zero and the discharge is neutrally buoyant. The equation is valid only close to the discharge, up to a distance corresponding to several (two to three) water depths. At longer distances, other factors are of increasing importance in jet mixing and must be included.[2]

A number of initial mixing models have been developed for the analysis of dilution of submerged discharges in the coastal water and are available from the U.S. EPA.[2]

B. AMBIENT TURBULENCE MIXING

Fischer et al.[3] presented the following two-dimensional mass transport model for ambient mixing in rivers with a discharge from the middle of the river channel:

$$C(x,y) = \frac{M}{hu(4\pi D_y x/u)^{1/2}} \exp\left[\frac{-y^2 u}{4D_y x}\right] \tag{8.1}$$

where

C = concentration at any given location
M = mass discharged/unit time
u = average velocity in the river
D_y = dispersion coefficient across the river
x = distance downstream from the diffusers
y = distance in lateral direction
h = average depth in the river

It should be pointed out that tidal rivers and estuaries are under tidal influence, resulting in longitudinal dispersion (spread) of the effluent. Thus, Equation 8.1 must be modified to incorporate the longitudinal spread along the estuary.

A two-dimensional depth-averaged, tidally-averaged advection-dispersion equation, which has been rigorously derived by Hamrick and Neilson,[5] is

$$\frac{\partial C}{\partial t} + u\frac{\partial C}{\partial x} + v\frac{\partial C}{\partial y} = \frac{1}{h}\frac{\partial}{\partial x}\left(hD_{xx}\frac{\partial C}{\partial x} + hD_{xy}\frac{\partial C}{\partial y}\right) + \frac{1}{h}\frac{\partial}{\partial y}\left(hD_{yx}\frac{\partial C}{\partial x} + hD_{yy}\frac{\partial C}{\partial y}\right) - K_d C \tag{8.2}$$

where

C = depth- and tidally-averaged concentration
u = depth- and tidally-averaged advective velocity in x direction
v = depth- and tidally-averaged advective velocity in y direction
D = dispersion coefficient tensor
K_d = first-order decay coefficient

The first solution to be examined is for a continuous point source discharge at $x = y = 0$ on the shoreline of a channel of width, B. In this case, x is positive in the direction of net flow (seaward) and y is positive across the channel toward the opposite shore at $y = B$. If the velocity field is unidirectional and the x coordinate is aligned in that direction, the transverse velocity v and the off-diagonal dispersion coefficients, D_{xy} and D_{yx}, may be set to zero. Furthermore, under steady-state conditions, Equation (8.2) becomes

$$hu\frac{\partial C}{\partial x} = \frac{\partial}{\partial x}\left(hD_x\frac{\partial C}{\partial x}\right) + \frac{\partial}{\partial y}\left(hD_y\frac{\partial C}{\partial y}\right) - hK_d C \tag{8.3}$$

where

$D_x = D_{xx}$ and $D_y = D_{yy}$

Hamrick and Neilson[5] provided detailed methods for determining the dispersion coefficients. In practice, tracer studies may be used to determine characteristics of the ambient mixing. An alternate method is to use dispersion coefficient values reported in the literature. Subsequent sensitivity analyses would substantiate and fine tune the dispersion coefficients.

Equation 8.3 has closed form analytical solutions only if the coefficients h, u, D_x, D_y, and K_d are constant.[5] Although this is seldom true for actual situations, it is possible to choose representative values that will give reasonable results in the immediate vicinity of the outfall.

The solution is

$$C = \frac{M}{\pi h (D_x D_y)^{1/2}} \exp\left[\frac{u}{(4K_d D_x)^{1/2}} \left(\frac{K_d}{D_x}\right)^{1/2} x\right]$$
$$\sum_{i=-\infty}^{\infty} K_0 \left[\left(1 + \frac{u^2}{4K_d D_x}\right)^{1/2} \left(\frac{K_d x^2}{D_x} + \frac{K_d}{D_y}(y + 2iB)^2\right)^{1/2}\right] \tag{8.4}$$

where

K_0 = the modified Bessel function of the second kind of order zero
B = depth-averaged width

If the estuary channel is sufficiently wide satisfying

$$B^2 >>> (D_y/K_d) \text{ or } B^2 > 300(D_y/K_d)$$

the series in Equation 8.4 may be truncated at the $i = 0$ term. Further, for conservative substances, $K_d = 0$, Equation 8.4 becomes:

$$C = \frac{M}{\pi h (D_x D_y)^{1/2}} \exp\left(\frac{ux}{2D_x}\right) K_0 \left[\frac{u}{2D_x^{1/2}} \left(\frac{x^2}{D_x} + \frac{y^2}{D_y}\right)^{1/2}\right] \tag{8.5}$$

Equation 8.5 represents a modification of Equation 8.1 by incorporating the longitudinal dispersion in a tidal river for far-field mixing. Equation 8.5 can be used to explicitly calculate the concentration at location given by (x,y).

Note that applying Equation 8.5 requires approximating the actual site conditions with idealized geometry, topography, and current fields. While this results in simple analytical solutions to the transport equations, the uncertainty must be incorporated into the dispersion coefficients. Thus, the development of more complex models allowing variations in geometry, topography, and current is needed. One of the methodologies to accommodate spatial variation of these factors is solving Equation 8.3 with a box model (such as HAR03 in Chapter 3) for mass transport. However, the assignment of spatially variable mass transport coefficients would still require a dye dispersion study in the field.

III. CASE STUDY — TOXICITY MIXING ZONE IN JAMES ESTUARY

The existing wastewater outfall of the Falling Creek wastewater treatment plant (WWTP) discharges the effluent into Falling Creek, which in turn enters the James Estuary (Figure 1). A new outfall has been designed to discharge the effluent directly into the James River. The proposed outfall is a 36-inch pipe on the western bank of the river. The acute impact of the discharge on the aquatic community in the vicinity of the outfall, is a primary concern.

The Virginia State Water Control Board (SWCB) originally proposed an acute whole effluent toxicity (WET) limit of 1 TU_a at the end of the outfall for the Falling Creek WWTP, i.e., not allowing any dilution. (See Section III.C.1 for the definition of acute toxicity unit, TU_a). The analysis presented in the following sections provides data and information to evaluate this limit.

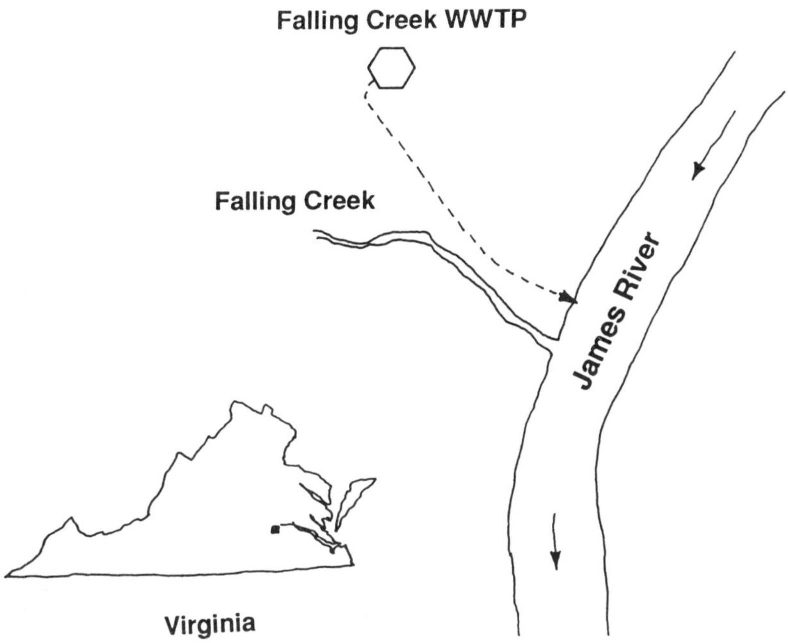

FIGURE 1. Falling Creek POTW discharging into the James Estuary.

This analysis addresses the important question: for the proposed acute toxicity limit, would the CMC to protect against acute or lethal effects be met in the ambient water?

A. WATER QUALITY STANDARDS

1. Overall Mixing Zone

In the recent amendments to the Water Quality Standards proposed by the SWCB, Section VR680-21-01.2.C allows mixing zones. However, no mixing zone established by the SWCB shall

- Interfere with passing or drifting aquatic organisms
- Cause acute lethality to passing or drifting aquatic organisms
- Be used for, or considered as, a substitute for minimum treatment technology required by the Clean Water Act and other applicable State and Federal laws
- Constitute more than one half of the width of the receiving watercourse nor constitute more than one third of the area of any cross section of the receiving watercourse
- Extend downstream at any time a distance more than five times the width of the receiving watercourse at the point of discharge

2. Zone of Initial Dilution

Further, an allocated impact zone may be allowed within a mixing zone. This zone is the area of initial dilution of the effluent with the receiving water where the concentration of the effluent will be its greatest in the water column. Mixing within these allocated impact zones shall be as quick as practical and shall be sized to prevent lethality to passing aquatic organisms. Mixing zones shall be determined such that acute standards are met outside the allocated impact zone and chronic standards are met at the edge of the mixing zone.

The U.S. EPA has developed guidelines to prevent the passing organisms from being exposed to lethal effects. Lethality to passing organisms can be prevented in the mixing zone in one of four ways according to the EPA.[2] The first method is to prohibit concentrations in excess of the CMC in the pipe itself, as measured directly at the end of the pipe.

The second method is to use high-velocity discharge with an initial velocity of 3 m/sec, or more, together with a mixing zone spatial limitation of 50 times the discharge length scale in any direction. If acute criteria (CMC derived from 48- to 96-h exposure tests) are met throughout the mixing zone, no lethality should result from temporary passage through the mixing zone. If acute criteria are exceeded no more than a few minutes in a parcel of water leaving an outfall, this likewise assures no lethality to passing organisms.

The third alternative is not to use a high-velocity discharge. Rather the discharger should provide data to the state regulatory agency showing that the most restrictive of the following conditions are met:

- The CMC should be met within 10% of the distance from the edge of the outfall structure to the edge of the regulatory mixing zone in any spatial direction.
- The CMC should be met within a distance of 50 times the discharge length scale in any spatial direction.
- The CMC should be met within a distance of five times the local water depth in any horizontal direction from any discharge outlet.

A fourth alternative is for the discharger to provide data to the state regulatory agency showing that a drifting organism would not be exposed to one-hour average concentrations exceeding the CMC, or would not receive harmful exposure when evaluated by other valid toxicological analysis. That is, if a full analysis of concentrations and hydraulic residence times within the mixing zone indicates that organisms drifting through the plume along the path of maximum exposure would not be exposed to concentrations exceeding the acute criteria when averaged over the one-hour (or appropriate site-specific) averaging period for acute criteria, then lethality to swimming or drifting organisms ordinarily should not be expected.[2]

The interpretation of the above alternatives suggests that no mixing zone calculations are needed under alternatives 1, 2, and 4. The first alternative is most stringent, requiring that CMC be met at the end of the pipe. The second alternative requires a very high discharge velocity (3 m/sec), which normally is not available with most gravity flows of wastewater. Further, under the 1Q10 low flow condition, the proposed outfall is significantly above the river water level at the discharge site, thus providing insignificant momentum-induced mixing.

The fourth method simply states that if the one-hour test results in no lethality, then the discharger is not required to have a toxicity permit. In general, few discharge outfalls fall into these three categories. Rather, most discharge outfalls are surface discharges with small or little momentum for initial mixing. As such, the third alternative is usually the situations encountered most frequently in practice. It means that mixing zone calculations are required if the regulations allow mixing zones.

B. STUDY APPROACH

Data from the study area have suggested that the first two methods of preventing lethality are not applicable to the Falling Creek WWTP outfall. The current design of the proposed outfall structure does not offer high-velocity discharge. In fact, the discharge velocity is much smaller than 3 m/sec. At the time of the study, there was no data on the one-hour test available. The study effort is therefore focused on the third alternative, providing data and information to the state to demonstrate that turbulence-induced mixing would be sufficient for meeting the CMC of 0.3 TU_a at the edge of the zone of initial dilution (or allocated impact zone).

A conservative approach of analysis was formulated for this analysis. That is, the near-field mixing modeling would be bypassed at the present time, thus neglecting discharge-induced mixing. As such, no credit is given for the momentum-induced mixing; mixing between the effluent and the river water is achieved only by turbulent mixing in the ambient water. Further, the effluent being modeled is assumed to be a conservative substance, given the spatial and temporal scales associated with the mixing zone.

To assist the analysis of the ambient induced mixing zone, dye dispersion studies are usually used to calibrate the mixing characteristics in the receiving water. In the field work, the Rhodamine WT dye is released with the effluent, which in turn is discharged into the receiving water. However, such a field study is not appropriate for a nonexisting outfall. Thus, literature data on the mixing coefficients in tidal rivers was used instead. Further, results from recent hydrodynamic modeling of the James Estuary by the Virginia Institute of Marine Science (VIMS) was used to develop the dispersion coefficients for the study area. The calculations were substantiated by model sensitivity analyses of the dispersion coefficients.

C. DATA ANALYSIS

1. Outfall Design and Effluent Characteristics

The proposed outfall has a pipe diameter of 36 in. The design flow of the Falling Creek WWTP is 10 mgd at the present time and this flow is used in the mixing zone calculation.

A toxicity unit is used to represent the toxicity level of the effluent. The acute toxicity unit is defined as:

$$TU_a = \frac{100}{LC_{50}} \tag{8.6}$$

where LC_{50} is the percent effluent that causes 50% of the organisms to die by the end of the acute exposure period. For example, an effluent with an acute toxicity in terms of LC_{50} of a 5% effluent is an effluent containing 20 TU_as. In this study, SWCB had proposed a whole effluent toxicity limit of 1 TU_a for the Falling Creek WWTP.

2. Completely Mixed Concentration

At the 1Q10 flow of 605 *cfs* and the effluent flow of 10 mgd (= 15.47 *cfs*), the maximum dilution is: (605 + 15.5)/15.5 = 40. That is, when the effluent is completely mixed with the river flow, the maximum dilution ratio would be about 40. The receiving water concentration for acute toxicity for comparison with the CMC is calculated to be:

$$C = \frac{1.0TU_a}{40} = 0.025\ TU_a$$

3. Dimensions of the Allocated Impact Zone

The dimensions of the allocated impact zone within which the CMC is met depends on the size of the regulatory mixing zone as specified in the State Water Quality Standards. In this analysis, they are calculated in the following steps:

1. The length of the regulatory mixing zone = 2700 ft (5 times the river width)
2. The width of the regulatory mixing zone = 270 ft (one half of the river width)
3. The CMC should be met within 10% of the lateral distance from the edge of the outfall structure to the edge of the mixing zone = 27 ft (10% of 270 ft)
4. The CMC should be met within a distance of 50 times the discharge length scale in any spatial direction = 150 ft (50 times the outfall pipe diameter of 3 ft)
5. The CMC should be met within a distance of five times the local water depth = 125 ft (5 times 25 ft)

Based on the above limitations, the size of the allocated impact zone is 27 ft (in the lateral direction) by 250 ft (in the longitudinal direction) for a shore discharge. As such, the CMC should be met at the edge of this zone.

4. Hydraulic Geometry and Ambient Mixing Coefficients

Figure 1 also shows the cross-sectional area (= 13,345 ft^2) in the study area under mean tide conditions. The average depth is 24.9 ft. Under the 1Q10 low flow of 605 *cfs* in the study area,

TABLE 1
Longitudinal and Lateral Dispersion Coefficients[a]

Site	Receiving water	D_x (ft²/s)	D_y (ft²/s)
Garrett's Marina	Rappahannock River	247	0.32
South Hill Banks Marina	Rappahannock River	247	0.32
Ingram Bay Marina	Ingram Bay	0.014	0.002
Cranes Creek	Ingram Bay	0.32	0.007
A.C. Fisher Marina	Cranes Creek	0.36	0.006
James River STP	James Estuary	172	0.12
York River STP	York Estuary	150	0.6

[a] From Hamrick and Neilson.[5]

TABLE 2
Summary of Lateral Dispersion Coefficients

Data source	River width (ft)	D_y (ft²/sec)
Yotsukura and Cobb[7]		
Missouri River near Blair	600	1.087
Beltaos[8]		
Athabasca below Ft. McMurray	1220	0.990
Beltaos[9]		
Bow River at Calgary	340	0.914

the average velocity is 0.045 ft/sec. Longitudinal and lateral dispersion coefficient values need to be assigned. To assist the selection, a range of longitudinal and lateral dispersion coefficients used by Hamrick and Neilson[5] for several marinas along the James River Estuary is listed in Table 1. Further, results from the hydrodynamic model of the James Estuary by Hamrick[6] were used to derive the dispersion coefficients for the study area: $D_x = 10$ ft²/sec and $D_y = 1.0$ ft²/sec. The longitudinal dispersion coefficient value selected is much smaller than those for the two STPs in the James River in Table 1. It should be pointed out that both the James Estuary and York Estuary sites in Table 1 are close to the mouths of these two rivers, subject to significant tidal actions. While the receiving water in the Falling Creek area is tidal, the tidal influence is diminished significantly, resulting in small longitudinal mixing. Another component in the longitudinal dispersion coefficient is the compensation from spatial averaging. In the two-dimensional model such as Equation 8.2, no lateral averaging is allowed. The only spatial averaging is in the vertical direction. The study area is located very close to the fall line and the vertical gradients of horizontal velocity is relatively small, somewhat similar to the vertical profile usually observed in a rivering system. As such, the second component contributing to longitudinal dispersion is also small in the study area.

Although the lateral dispersion coefficient of 1.0 ft²/sec selected for this analysis is slightly greater than most of the reported values in Table 1, it is consistent with some other literature values of lateral mixing coefficients in rivers as listed in Table 2.

D. MODEL APPLICATION

1. Evaluation of Acute Toxicity Limit of One TU_a

Equation 8.5 is applied with the following data:

- Total wastewater flow = 10 mgd
- Effluent acute toxicity = 1.0 TU_a
- River velocity = 0.045 ft/sec

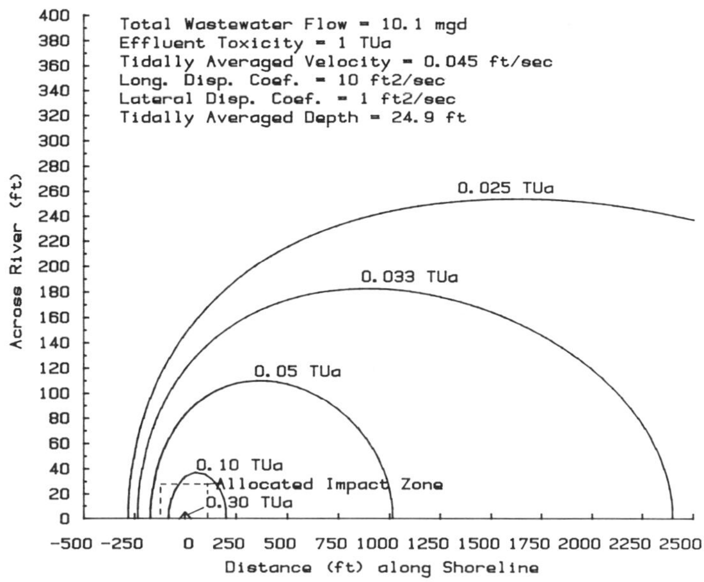

FIGURE 2. Model calculated isopleths with effluent toxicity = 1 TU_a.

- Longitudinal dispersion coefficient = 10 ft²/sec
- Lateral dispersion coefficient = 1.0 ft²/sec
- River depth = 24.9 ft

The model results are shown in Figure 2 in which five isopleth toxicity contours are displayed. Contour No. 1 has a toxicity of 0.3 TU_a and contour No. 5 represents the toxicity level of complete mixing (i.e., 0.025 TU_a). The other three isopleths are 0.033, 0.05, and 0.10 TU_a, respectively. The model results show that the 0.3 TU_a isopleth is within the allocated impact zone, thus meeting the water quality standard for whole effluent toxicity.

2. Model Sensitivity Analysis

Most of the model input parameters associated with Equation 8.2 are independently determined. Only the longitudinal and lateral coefficients are indirectly derived and are thereby, subject to certain degrees of uncertainty. Therefore, the model sensitivity analyses of these two parameters are conducted. First, the longitudinal dispersion coefficient, D_x, is varied from 5 to 20 ft²/sec. The model results indicate that the mixing zone calculations are not sensitive to this parameter.

An empirical equation to calculate the lateral dispersion coefficient from Fischer et al.[3] may be used to develop the range of the values for a model sensitivity analysis:

$$D_y = \phi h u^* \tag{8.7}$$

where

ϕ = an empirical constant ranging from 0.41 to 0.65
u* = shear velocity (ft/sec)

Using a shear velocity of 0.1 ft/sec, which is considered reasonable for the study area, and an average depth of 24.9 ft, one could calculate a range of D_y between 1.02 and 1.495 ft²/sec. Fischer[10] reported that higher values of ϕ are usually found near the banks of rivers as in this case. The higher D_y value offers more rapid mixing between the effluent and the ambient water, would generate a 0.3 TU_a isopleth closer to the discharge point, when comparing with that in Figure 2.

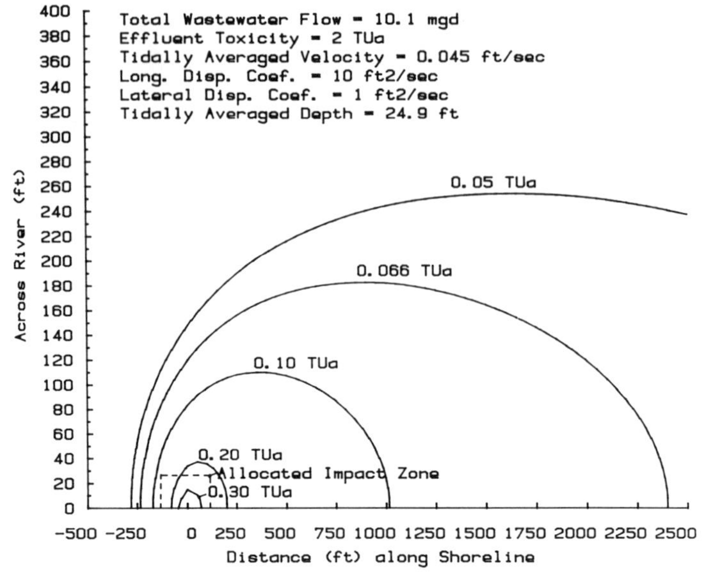

FIGURE 3. Model calculated isopleths with effluent toxicity = 2 TU_a.

[Thus, the results in Figure 2, which are based on the low end value of D_y (1.0 ft^2/sec), are again on the conservative side!]

3. Model Prediction

The ambient mixing model was then used to evaluate a higher effluent toxicity limit, such as 2 TU_a. The results of the model calculations are shown in Figure 3, using the low end value of D_y (1.0 ft^2/sec). It is seen that the 0.3 TU_a isopleth is still within the allocated impact zone, suggesting that a 50% dilution of the effluent may be allowed.

IV. MASS TRANSPORT CALIBRATION USING DYE DISPERSION DATA

The modeling framework described in Section III is limited to constant hydraulic geometry and mass transport coefficients such as tidally averaged velocity, depth, longitudinal, and lateral dispersion coefficients. For more complicated geometry, an advective-dispersion mass transport model is needed to determine the transport coefficients. One of the commonly used approaches is using dye dispersion data to calibrate the mass transport coefficients in a box model configuration.

A tracer or dye study can be used to determine the areal extent of mixing in an estuary, the boundary where the effluent has completely mixed with the ambient water, and the dilution that results from the mixing. Obviously, if the outfall is not yet in operation, it is impossible to determine discharge-induced mixing by tracer studies. Tracer studies can be used in these situations to determine characteristics of the ambient mixing. For ambient mixing studies, the tracer release can be either instantaneous or continuous. Instantaneous releases are used frequently to measure longitudinal dispersion, but can also be used to determine lateral mixing in rivers and lateral and vertical mixing in estuaries, bays, and lakes.[2] For water bodies with significant flow velocities, continuous releases of tracer are normally used to determine lateral and vertical mixing coefficients. Continuous releases can also be used to determine three-dimensional concentration isopleths for steady-state conditions. A number of references

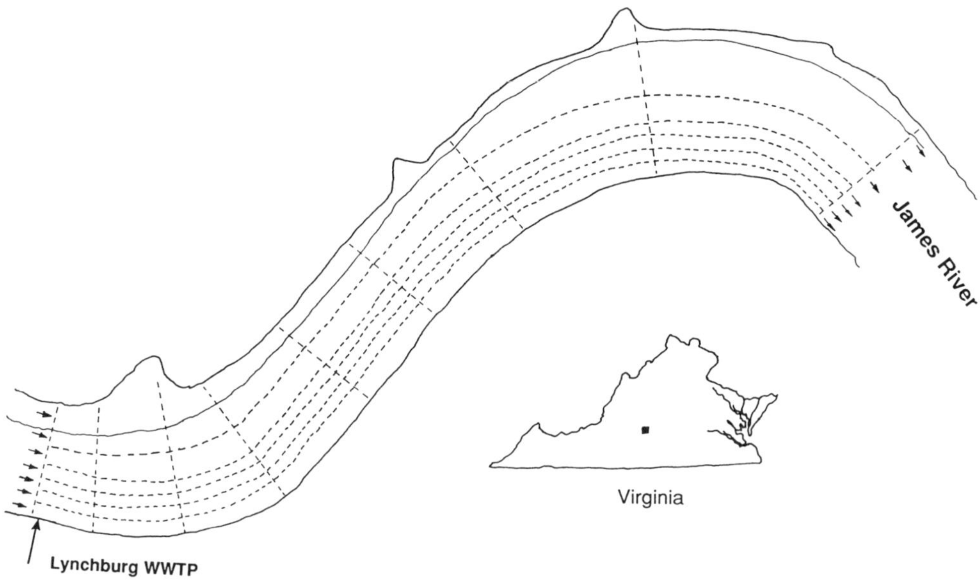

FIGURE 4. Model segmentation for mass transport model of the James River.

provide information concerning the design, conduct, and analysis of tracer studies for mixing analyses.[11,12]

A. STUDY AREA AND FIELD DATA COLLECTION

The City of Lynchburg's WWTP discharges its effluent into the James River in Virginia (Figure 4). The dye dispersion study was part of a modeling analysis to assess the mixing zone of total chlorine residual (TRC) in the effluent. To assist in the design and planning of the field investigations, a simple plume equation[1] was used:

$$C(x,y) = \frac{M}{hu(4\pi D_y x / u)^{0.5}} \exp\left(\frac{-y^2 u}{4D_y x}\right) \tag{8.8}$$

Note that Equation 8.8 is for riverine systems with longitudinal dispersion and is somewhat similar to Equation 8.5 for estuarine systems.

Nine sampling transects were established. One transect upstream of the outfall served as the background condition and the eight downstream transects were used to establish the mixing characteristics. A continuous release of Rhodamine WT dye was administered at the plant. The river flow on the day of the field survey was 1525 *cfs* and the wastewater flow was 22 mgd.

B. MODEL DEVELOPMENT AND CALIBRATION

A box mass transport model was developed to calculate the dye concentrations with a total of 48 segments (Figure 4). The modeling framework HAR03[13] was used for this analysis (see Chapter 3 for the discussions and examples of using this modeling framework). The most important mass transport coefficients in this modeling analysis are the lateral dispersion coefficients, D_y, which are spatially variable. Preliminary estimates of D_y were obtained using Equation 8.7. Subsequent model calibration analyses determined the spatially variable D_y values by matching the calculated dye concentrations with the data.

The calculated and measured dye concentrations in the lateral direction are shown in Figure 5 for eight transects across the river. As indicated, the model results match the measured dye

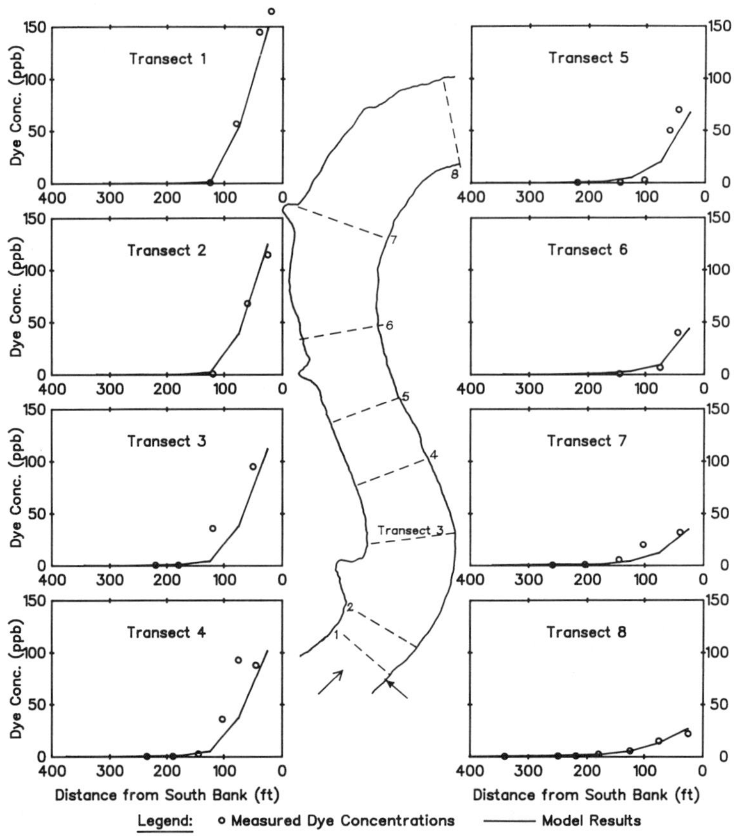

FIGURE 5. Calculated and measured dye concentrations in transects across the James River.

concentrations closely following a series of model calibration and sensitivity analysis runs. Figure 5 also shows that high concentrations (above 100 ppb) are attached to the south bank, where the effluent was discharged, from Transect 1 to Transect 4. In the meantime, dispersion in the lateral direction progresses. Both the data and model results show that the dye has reached 300 ft across the south bank at Transect 8, which is located about 3500 ft downstream of the outfall.

Figure 6 shows the longitudinal profiles of dye concentrations in the three tubes closest to the south bank. Again, the model results match the data very well. The calibrated lateral dispersion coefficients range from 0.44 to 1.22 ft^2/sec, which are realistic and within the literature reported ranges for their values.[14]

V. SUMMARY

The concept of mixing zone and near-field and far-field mixing are presented in this chapter. In addition, the mixing zone regulations designed to protect the aquatic life against lethality is discussed. To quantify the mixing zone in estuaries, a simplified two-dimensional, steady-state model for far-field mixing is presented. The case study demonstrates a successful application of the model to the James River Estuary in Virginia.

One of the limitations associated with this model is using constant hydraulic geometry in the calculation. For more complicated situations, a mass transport model based on the box model approach may be used in conjunction with a dye study in the field to independently determine

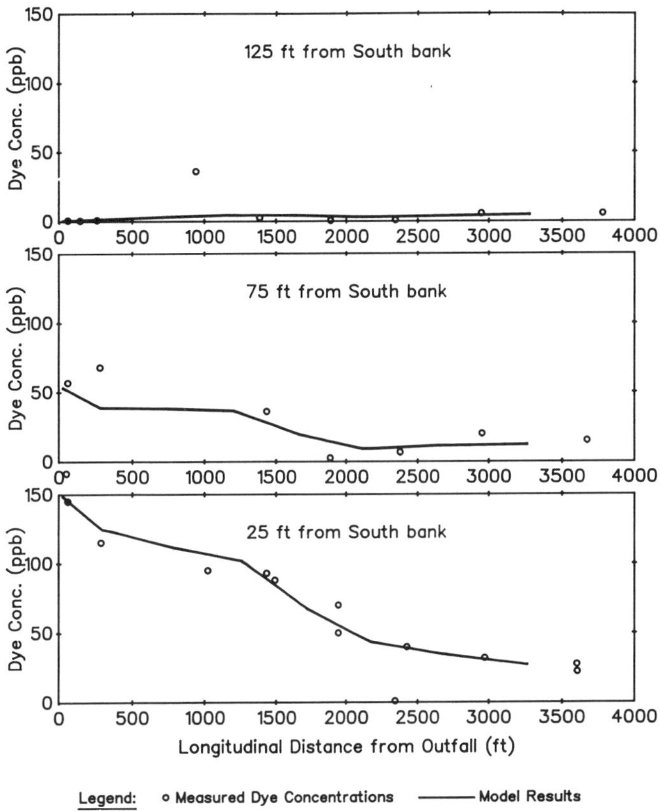

FIGURE 6. Calculated and measured dye concentrations in longitudinal direction along the James River.

the spatially variable mixing coefficients. Another case study illustrating the use of dye dispersion field data to calibrate the mixing coefficients was presented.

A field dye dispersion study is an expensive and time-consuming study to determine the mass transport coefficients, particularly the longitudinal and vertical dispersion coefficients. When using these models as a screening tool, literature values may be used as preliminary estimates of these coefficients.[14,15] Subsequent model sensitivity analyses will provide additional insight as to the significance of the mixing characteristics. This approach is particularly useful for an order-of-magnitude analysis prior to any field dye dispersion study.

It should be pointed out that many exiting outfalls, either municipal or industrial wastewaters, lack momentum to provide any significant initial dilution. In fact, this situation becomes more pronounced when addressing the acute toxicity under the 1Q10 low flow conditions. That is, the receiving water surface elevation would be so low under this low flow condition that most of the outfall pipes are located above the receiving water surface, making the initial dilution momentum almost zero. Many existing outfalls deliver the wastewater to the receiving water by gravity flow, also lacking the high velocity (e.g., above 3 m/sec) as a jet. Under this circumstance, turbulence mixing provided by the ambient water is the only mixing mechanism available to the effluent. The methodology presented in this chapter is suitable to these conditions encountered in the field.

REFERENCES

1. **Neely, W. B.,** The definition and use of mixing zones, *Environ. Sci. Technol.*, 16, 518A, 1982.
2. **U.S. EPA,** *Technical Support Document for Water Quality-based Toxics Control*, EPA/505/2-90-001, 1991, 77.
3. **Fischer, H. B., List, E. J., Koh, R. C. Y., Imberger, J., and Brooks, N. H.,** *Mixing in Inland and Coastal Waters*, Academic Press, Inc., New York, 1979.
4. **Holley, E. R. and Jirka, G.H.,** Mixing in Rivers, Army Engineers Waterways Experiment Station, technical report E-86-11, Vicksburg, MS, 1986.
5. **Hamrick, J. M. and Neilson, B. J.,** *Determination of Maria Buffer Zones Using Simple Mixing and Transport Models*, report submitted by Virginia Institute of Marine Science for Virginia Department of Health, 1989, 68.
6. **Hamrick, J. M.,** "Long-term dispersion in unsteady skewed free surface flow," *Estuarine Coastal Shelf Sci.*," 23, 807, 1986.
7. **Yotsukura, N. and Cobb, E. D.,** *Transverse Diffusion of Solutes in Natural Streams*, U.S. Geological Survey professional paper 582-C, 1972.
8. **Beltaos, S.,** Mixing processes in natural streams, in *Transport Processes and River Modeling Workshop*, Canada Centre for Inland Waters, 1978.
9. **Beltaos, S.,** *Transverse Mixing in Natural Streams*, Transportation and Surface Water Eng. Div., Alberta Research Council, Report No. SWE-78/01, 1978.
10. **Fischer, H. B.,** Dispersion predictions in natural streams. *ASCE J. Sanit. Eng.*, 94, 927, 1968.
11. **Hubbard, E. F., Kilpatrick, F. A., Martens, L. A., and Wilson, J. F.,** Measurement of Time of Travel and Dispersion in Streams by Dye Tracing. Book 3, Application of Hydraulics, Chapter A9, U.S. Geological Survey, 1982, 44.
12. **Wilson, J. F., Cobb, E. D., and Kilpatrick, F. A.,** Fluorometric Procedures for Dye Tracing, Book 3, Application of Hydraulics, Chapter A12, U.S. Geological Survey, 1986, 34.
13. **Applied Technology and Engineering,** Mixing Zone Analysis of the James River at the Lynchburg Regional Wastewater Treatment Plant, report prepared for City of Lynchburg, VA, 1989, 35.
14. **Mills, W. B., Porcella, D. B., Ungs, M. J., Gherini, S. A., Summers, K. V., Mok, L., Rupp, G.L., and Bowie, G. L.,** Water Quality Assessmen t: A Screening Procedure for Toxic and Conventional Pollutants in Surface and Ground Water — Part I (Revised–1985), U.S. Environmental Protection Agency, EPA/600/6-85/002a, 1985, 6.
15. **Schnoor, J. L., Sato, C., McKechnie, D., and Sahoo, D.,** Processes, Coefficients, and Models for Simulating Toxic Organics and Heavy Metals in Surface Waters, U.S. Environmental Protection Agency, EPA/600/3-87/015, 1987, 2.

INDEX

N

O

P

V

W

Z